Genilson Lima Diniz
Reginaldo Gomes Nobre
Geovani Soares de Lima

Ecophysiology of passion-fruit plant under saline water and silicate fertilization

Genilson Lima Diniz
Reginaldo Gomes Nobre
Geovani Soares de Lima

Ecophysiology of passion-fruit plant under saline water and silicate fertilization

In the semi-arid conditions of Northeast Brazil

ScienciaScripts

Cover image: www.ingimage.com

This book is a translation from the original published under ISBN 978-620-3-46819-9.

Publisher:
Sciencia Scripts
is a trademark of
Dodo Books Indian Ocean Ltd., member of the OmniScriptum S.R.L Publishing group
str. A.Russo 15, of. 61, Chisinau-2068, Republic of Moldova Europe
Printed at: see last page
ISBN: 978-620-3-69755-1

CONTENTS

The yellow passion-fruit plant is considered to be one of the principal fruits with a high market value, which has aroused interest mainly for small and medium-sized producers in the semi-arid north-eastern region. However, in these regions it is often necessary to use water with high concentrations of salts for irrigation. For the use of these waters, many times it requires management strategies, the silicate fertilization that presents itself as an alternative of attenuation to this stress condition for the plants. In this context, the objective was to evaluate the cultivation of yellow passion-fruit irrigated with waters of different salinities associated with silicate fertilization. The experiment was carried out in two stages at the Federal University of Campina Grande, Pombal - PB. The first stage was carried out in a greenhouse environment, where different levels of salinity and doses of silicon were tested in the production of seedlings of yellow passion-fruit, with randomized block design, with a factorial scheme 5 x 5, whose treatments were constituted by five levels of electrical conductivity of the irrigation water (0.3; 1.0; 1.7; 2.4 and 3.1 dS m-1) and five doses of silicate fertilizer (0; 25; 50; 75 and 100 g of potassium silicate/plant) with four repetitions and two plants per plot, totalizing two hundred experimental units. The salinity in the irrigation water reduced the photosynthetic pigments and yield in photosystem II in yellow passion-fruit plants. Silicate fertilization at a dose of 75 g plant-1 attenuated the deleterious effect of salt stress on osmotic potential, chlorophyll b and carotenoids. Salt stress caused by conductivity water at 0.3 dS m-1 and above caused a decrease in the production of phytomass of passion-fruit at 60 days after sowing, with the dry phytomass of the leaves being the most sensitive variable. The production of passion-fruit seedlings with water salinity up to 3.1 dS m-1 obtained an acceptable Dickson quality index. The doses of silicate fertilization softened the effect of salt stress only for the root/head ratio of the plants of yellow giant passion-fruit. In the second trial, the experiment was developed in field conditions. The treatments consisted of five levels of electrical conductivity of the irrigation water (0.3; 1.0; 1.7; 2.4 and 3.1 dS m-1) associated with two doses of silicate fertilizer (150 and 300 g plant-1 of potassium silicate) with four repetitions. Salt stress caused a decrease in stomatal conductance, intercellular CO_2 concentration, transpiration, instantaneous water use efficiency and growth of passion-fruit plants. The synthesis of *chlorophyll a* was inhibited by the increase of the CEa from 0.3 dS m-1, however, the salinity of the water caused an increase in the contents of carotenoids and protoplasmatic content; the silicon had a significant effect in the conductance, photosynthesis, efficiency of water use, chlorophyll b, carotenoids and rate of relative growth of the diameter.

Passiflora edulis f. flavicarpa, salinity, silicon.

Abstract: The yellow passion fruit is considered one of the main fruit plants with high market value, which has aroused interest mainly for small and medium producers in the northeastern semiarid. However, in those regions it is often necessary to use waters with high concentrations of salts for irrigation. For the use of these waters, management strategies are often required, silicate fertilization, which presents itself as an alternative to mitigate this stress condition for plants. In this context, the objective was to evaluate the cultivation of yellow passion fruit irrigated with waters of different salinity associated to silicate fertilization. The experiment was carried out in 2 stages at Federal University of Campina Grande, Pombal - PB. The first stage was carried out in a greenhouse environment, where different levels of salinity and doses of silicon were tested in the production of yellow passion fruit seedlings, with a randomized block design, with a 5 x 5 factorial scheme, whose treatments consisted of five levels of electrical conductivity of irrigation water (0.3; 1.0; 1.7; 2.4 and 3.1 dS m-1) and five doses of silicate fertilization (0; 25; 50; 75 and 100g of potassium silicate / plant), with four replications and two plants per plot, totalizing two hundred experimental units. The salinity in the irrigation water reduced the photosynthetic pigments and the yield in photosystem II in yellow passion fruit plants. The silicate fertilization in the 75g plant-1 dose attenuated the deleterious effect of salt stress for the osmotic potential, chlorophyll b and carotenoids. The salt stress caused by the conductivity of water from 0.3 dS m-1 caused a decrease in the production of phytomass of passion fruit at 60 days after sowing, being the dry leaf phytomass highlighted as the most sensitive variable. The production of passion fruit seedlings with water salinity up to 3.1 dS m-1 obtained an acceptable Dickson quality index. The doses of silicate fertilizer softened the effect of saline stress only for the root / shoot ratio of giant yellow passion fruit plants. In the second trial, the experiment was carried out under field conditions. The experimental design was used in randomized blocks in a 5 x 2 factorial scheme, whose treatments consisted of five levels of electrical conductivity of irrigation water (0.3; 1.0; 1.7; 2.4 and 3.1 dS m-1) associated with two doses of silicate fertilization (150 and 300 g plant-1 of potassium silicate) with four replications. Saline stress provided a decrease in stomatal conductance, intercellular CO2 concentration, transpiration, instant water use efficiency and in the growth of passion fruit plants. The synthesis of chlorophyll a was inhibited by the increase in CEa from 0.3 dS m-1, however, the salinity of the water caused an increase in the levels of caratenoids and protoplasmic content; silicon had a significant effect on conductance, photosynthesis, water use efficiency, chlorophyll b, caratenoids and relative diameter growth rate.

Keywords: Passiflora edulis f. flavicarpa, salinity, silicon.

1 GENERAL INTRODUCTION

The yellow passion fruit (*Passiflora edulis* f. flavicarpa) is cultivated in almost all Brazilian territory, where Brazil stands out as the largest producer and consumer being the juice of this fruit the third most produced by Brazilian agroindustries (FREIRE et al., 2014). According to data from IBGE (2018) in the year 2017 the national production was 293,699.535 tons in an area of 34,672 establishments, highlighting Bahia as the main producer at the national level with 110,470.062 tons, Ceará in fourth place with 19,559.205 tons and Pernambuco occupying the sixth place with 8,168.605 tons.

In the state of Paraíba, production in 2017 was 5,334.137 tons with 1,138 establishments. The main producing municipalities were Santa Rita, Nova Floresta, Picuí, Pitimbu, Araruna, Cuité and Conde (BRASIL, 2017a).

The production of seedlings is a very important strategy for good productivity, because low quality seedlings will negatively influence the production of the crop (DINIZ; GUIMARÃES; LUZ, 2006). Thus, the planting of good quality seedlings influences the success of the implantation of the crop, providing control of the initial stand of plants. This reduces the need for some initial cultural treatments such as thinning, weeding, irrigation and spraying, providing greater homogeneity among the plants that are being cultivated (SALATA et al., 2011).

Despite the good indices of productivity of passion-fruit in the north-east region, its production depends on the use of irrigation, because of the dryness of the rains, high evapo-transpiration and high temperatures. In several cases, production is affected by the high levels of salts present in the surface and ground waters of this region. Salinity affects the availability of water to plants because of changes in the osmotic potential of the soil solution, reducing it to the point that the plant cannot easily extract the water (GHEYI; DIAS; LACERDA, 2010).

Ribeiro et al. (2014) using CEa levels ranging from 0.27 to 4.5 dS m-1 found that when salinity levels higher than 1.5 dS m-1 were used there were reductions in the growth and development of passion fruit.

However, crop tolerance to salinity can be determined as the ability of the plant to tolerate certain levels of salts, which is variable in relation to genotype, developmental stage, cationic nature of the water, stress intensity and soil and water management practices (SÁ et al., 2013; OLIVEIRA et al., 2015).

Among the strategies that may favour agricultural exploration using saline waters in irrigation, the management of water and fertilizers, among which silicon is mentioned, stands out. Despite the fact that silicon is not considered an essential element for plant growth, it has presented several benefits for crops, including an increase in the photosynthetic rate, a reduction in the toxic effect of some heavy metals, associated with an increase in the capacity for antioxidant defence when faced with water deficiency (GONG et al., 2005; HATTORI et al., 2005).

With the intention of mitigating the effects of salinity on crop development, several studies have been carried out using silicon (Si) as a salt stress attenuator. Ashraf et al. (2010) in a research conducted with sugarcane under salt stress observed that the addition of Si reduced the absorption and transport of Na+ to the aerial part and increased the absorption of K+, improving growth and consequently its production. Moussa and Galad (2015) in research on maize cultivars with treatment of 0 to 150 nM NaCl and 2.5 nM Si, found beneficial effects on Si addition, with Na+ uptake being inhibited by Si on transpiration flux. The application of Si increased the contents of N, P and K in maize plants causing them to obtain a better yield.

Although Si is an important element for inhibition of abiotic and biotic stresses, there are few reports in the literature about its use in the growth, physiology, production and post-harvest quality of passion-fruit. Considering the importance of passion-fruit culture in the Northeast region and the great need for the use of water with high levels of salt in this region, the development of research to identify a technique capable of alleviating salt stress is of extreme importance for the production of passion-fruit on a commercial scale.

2 THEORETICAL FRAMEWORK

2.1 The culture of the yellow passion-fruit

The yellow passion-fruit belongs to the class Magnoliopsida, order Malpigiales, family Passifloraceae and the genus Passiflora, it is a semi-perennial climbing plant with its centre of origin in Tropical America, and may reach 5 to 10 m in length, with more than 150 species native to Brazil, being cultivated more frequently in countries with tropical and subtropical climates (FALEIRO; FARIAS NETO; RIBEIRO JÚNIOR, 2008) gaining prominence as one of the most important fruit trees representing the fruit production chain at the small and medium producer level.

The root system of the plants is axial or pivotal, where 70% to 85% of the secondary roots are distributed up to the first 40 cm of soil depth (SOUSA et al., 2002). Its stem has a circular characteristic, woody and quite lignified, decreasing the lignin content as it approaches the apex of the plant (GRECO, 2014).

Its flowers are hermaphrodite, isolated or in pairs, located in the axils of the leaves and grouped in racemose or fasciculate inflorescences, which stand out because they are beautiful and vigorous flowers that attract people's attention. Its fruit is a berry that is attached by a peduncle with sometimes lignified skin (GRECO, 2014).

It stands out for its great importance, due to the sensory and pharmaco-therapeutic characteristics of the fruit and its good acceptance on the domestic market, thus representing an economic alternative with a fast return on the capital that was invested, since it is widely cultivated on small and medium-sized properties (GONDIM et al., 2006; PIRES; MATA, 2004; PIRES, 2008).

The passion-fruit plant in Brazil is expanding in the fruit market, presenting promising prospects for producers in various regions that use this fruit for production (MENDONÇA, 2011). This species is well accepted in family farming in the sertão of

Paraibano because the soil and climate conditions of the region are favourable for its development.

The passion fruit is used mainly for consumption *in natura* and manufacture of juices that can be sold on the domestic market and also exported. In 2017 Brazil obtained a production of 293,699.535 tons with 34,672 establishments, among the main yellow passion fruit producing states we can highlight Bahia as the main producer, followed by Minas Gerais, Santa Catarina, Ceará and Goiás (BRASIL, 2017b).

The phase of production of seedlings of fruit trees in general constitutes one of the most limiting phases for the implantation of the plantations, since seedlings of good quality influence the initial growth, the physiological aspects and development of the culture (LENZA, 2009). The passion-fruit plant has a short cycle, with production starting between 6 and 9 months after planting (SILVA, 2012) and a cultural cycle of around two years. In producing regions, this culture demands a constant renewal of the planted areas by the producer, forcing him to produce or acquire high quality seedlings to have good productivity rates.

2.2 Saline water as an alternative for irrigated crops in the semi-arid region

Food production can be reduced due to various abiotic stresses, among these stresses the salinity of water and/or soil allows the highest rates of reduction in food production. About 20% of agricultural land in the world is affected by salinity problems (RASSOL et al., 2013). These stresses may originate naturally related to pedogenesis or by anthropogenic action where it causes greater economic impact (JAYAKANNAN et al., 2015).

The salinization of anthropogenic origin is related to the inadequate management of irrigation in agricultural areas, mainly due to the use of artesian well water, where most of them have high salt contents of 11.802 dS m $^{-1}$ (QADOS; MOFTAH, 2015).

The term salinity can be defined as the existence of soluble salts in the soil that may harm the development and production of crops. The main salinity-related ions are the cations sodium (Na+), calcium (Ca2+), magnesium (Mg^{2+}), potassium (k+) and the anions chloride (Cl-), sulphate (SO42-) bicarbonate (HCO3-), carbonate (CO32-) and nitrate (NO3-) (YADAV et al., 2011).

The scarcity of water all over the world makes the management of irrigation gaining importance in the agronomic field. The growing demand for food and the limited supply of agricultural areas make it possible to use soils with salinity problems for agricultural exploration. Saline soils can generally be found in arid and semi-arid regions and present electrical conductivity (EC) values equal to or greater than 4.0 dS m-1 (MUNNS; TESTER, 2008).

The Northeast region is characterized by low rainfall, poorly distributed rainfall and high water loss by evaporation, presenting more than 60% of arid climate areas (MEDEIROS et al., 2012), contributing to an increase in the concentration of salts in groundwater and artesian wells, thus compromising the quality of irrigation water.

The use of saline water can contribute to an increase in the income of some families through agriculture, so that areas that did not have agricultural development as their main source of income can be reactivated. The availability of water for irrigation can be increased through the judicious use of saline water and recycling of drainage water; water currently considered inappropriate for irrigation can be used without major long-term consequences for crops and soils, provided that adequate cultural management practices and salinity-tolerant crops are adopted (RHOADES; KANDIAH; MASHALI, 1992; STEPPUHN, 2001).

In situations of salt stress the osmotic and ionic effects influence promoting alterations in the metabolic activities of the cells and in the process of cell elongation, directly affecting the growth of the plant, and depending on the intensity of the stress lead to the death of the plant (SAÍRAM; TYAGI, 2004; TAIZ; ZEIGER, 2013). Excess salts in plants can induce damage to proteins, lipids and nucleic acids, also occurring changes in the processes of photosynthesis and respiration affecting the growth and development of plants (BALAKHNINA; BORKOWSKA, 2013; MANAI et al., 2014).

2.3 Tolerance of passion-fruit to salinity

Cultivated plants are classified according to the degenerative action of salts on germination, growth and production. They are classified as sensitive when the electrical conductivity of the saturation layer of the soil is greater than 1.3 dS m-1, moderately sensitive when the salinity is greater than 3.0 dS m-1, moderately tolerant when the salinity is greater than 6.0 dS m-1 and tolerant or resistant when the salinity reaches levels greater than 10.0 dS m-1. The yellow passion-fruit is sensitive to salinity (AYERS; WESTCOT, 1999).

Sousa et al. (2006) evaluated passion-fruit plants subjected to irrigation with saline water greater than 1.5 dS m-1 and found that there was inhibition in the germination process, initial growth in height, number of leaves, leaf area and production of biomass of the roots and aboveground parts of the plants.

A study by Araújo et al. (2013) found that increasing the salinity of irrigation water reduced the percentage of emergence in yellow passion-fruit plants. A study by Oliveira et al. (2015) evaluating the effects of water salinity (ECa ranging from 0.8 to 3.6 dS m-1) showed that at 1.5 dS m-1 there was an increase in the concentration of sodium salts in the substrate, exerting phytotoxicity on the passion-fruit plants, with hormonal and ionic alterations affecting plant growth.

Freire et al. (2014) studying the gas exchange of passion-fruit irrigated with saline water (4.5 dS m-1) found that salt stress inhibits photosystem II photochemical activity and net photosynthesis of the plants. A study by Dias et al. (2011) evaluating the production of yellow passion-fruit, cultivated with bio-fertilizer irrigated with water of increasing electrical conductivity, observed that increasing the concentration of salts in irrigation water drastically reduces the production and number of fruits per plant of yellow passion-fruit.

Viana et al. (2012) studying the effect of salinity on the production of yellow passion-fruit concluded that water with an electrical conductivity greater than 2.5 dS m-1 significantly affects the production of yellow passion-fruit. Dias et al. (2012) found that the increase in the salinity concentration of irrigation water caused losses in the production and physical qualities of the fruit of yellow passion-fruit, being more drastic at conductivity levels greater than 2.5 dS m-1.

2.4 Silicon as a salt stress reliever

Although silicon (Si) is not considered an essential element for plants, it presents several benefits for various agriculturally cultivated species, among them: increase and maintenance of the photosynthetic rate, increase of plant stomatal conductance, decrease of the toxic effect of Mn, Fe and other heavy metals, in addition to minimizing the negative effects that salinity promotes on plants, with this the use of Si is one of the strategies that has been expanding in recent decades (SIVANESAN et al., 2011).

Si is absorbed by the roots in the form of monosilicic acid, is transported via the xylem, by transpiration flow, or actively, using specific transport proteins, this happens when the plant is subjected to stress or conditions of attack by pests and diseases (OLIVEIRA et al., 2009). Si is deposited in the cell walls, cell lumen and intercellular spaces of the root and stem (SANGSTER; HODSON; PARRY, 2001).

Plants in general contain Si in their tissues, and this concentration in the aerial part varies greatly according to each species (0.1 to 10% of Si in dry weight), showing an unequal distribution in plants (LIANG et al., 2015). It is divided into three groups taking into account their ability to absorb and accumulate Si in plant organs.

The species that are silicon accumulators, in a general way, the monocotyledons (Gramineae), have the active process of silicon absorption, having a Si content above 10.0 Kg-1 in dry matter, the species that are considered non-accumulators (legumes), species, which absorb silicon as a result of transpiration flowing more slowly than the absorption of water, with a content of less than 5.0 Kg-1 in dry matter and intermediate species, where the accumulation of silicon is absorbed by the roots at the same rate as the absorption of water, with a content of less than 10 g Kg-1 (CORNELIS et al., 2011).

The supply of Si to plants has been studied, recently, from the liberation of the use of potassium silicate (K_2SiO_3) as a fertilizer, where it is a source for the supply of this element, currently already being used in agriculture, with the purpose of mitigating the deleterious effects caused by stress, both biotic and abiotic (FREITAS, 2011).

Several studies are being developed to clarify all the mechanisms by which silicon contributes to the tolerance of plants when subjected to salt stress. In this way some propositions have gained more acceptance by the scientific community, the first of them is that Si decreases the concentration of salts in the plant due to the reduction of transpiration by the accumulation of the element in the leaf, the second is that Si decreases the transport of (Na^+) in the roots and the third proposition is that the element has physiological functions that cause it to increase the antioxidant metabolism (SHI et al., 2013).

Ferraz et al. (2015) in a study conducted with castor oil plant subjected to salt stress and fertilization with silicon, concluded that the increase in Si concentrations raised the values of total chlorophyll independent of salt levels.

Recent studies have shown beneficial effects of Si on the growth of many plant species (canola, soybean, wheat, sorghum, tomato and corn) subjected to salt stress conditions, increasing leaf area, chlorophyll content and improvements in chloroplast structure, which provided increased photosynthetic activity (HASHEMI; ABDOLZADEH; SADEGHIPOUR, 2010; LEE et al, 2010; TAHIR et al., 2012; BAE et al., 2012; HAGHIGHI; PESSARAKLI, 2013; ROHANIPOOR et al., 2013).

However, there are no reports in the literature about the use of silicon in passion-fruit cultivation, so it is of fundamental importance to carry out studies in order to find alternatives that contribute to the mitigation of the problems caused by salinity in the crop.

BIBLIOGRAPHIC REFERENCES

ASHRAF, M; RAHMAUTULLAH, M; AFZAL, R; AHMED, F; MUJEEB, A; SARWAr L. Alleviation of deterimental effects of NaCl by silicion nutrition in salt-sensitive and salt-tolerant genotypes of sugarcane (*Saccharum officinarum* L.). **Plant and Soil**, v. 326, n. 2 p. 326-381, 2010.

AYERS, R. S.; WESTCOT, D. W. **A qualidade da água na agricultura**. Campina Grande: Universidade Federal da Paraíba, 1999. 153p.

ARAÚJO, W. L. et al. Produção de mudas de maracujazeiro-amarelo irrigadas com água salina. **Agropecuária Científica no Semiárido,** v. 9, n.4, p. 15-19, 2013.

BAE E. J.; LEE, K. S.; HUTH, M. R.; LIM, C. S. Silicon significantly alleviates the growth inhibitory effects of NaCl in salt-sensitive 'Perfection' and 'Midnight' Kentucky bluegrass (*Poa pratensis* L.). **Horticulture, Environment, and Biotechnology**, v. 53, n. 6, p. 477-483, 2012.

BALAKHNINA, T.; BORKOWSKA, A. Effects of silicon on plant resistance to environmental stresses: review. **Institute of Agrophysics**, v. 27, n. 2, p. 225-232, 2013.

CORNELIS, J.T.; DELVAUX, B.; GEORG, R.B. Tracing the origin of dissolved silicon transferred from various soil-plant systems towards rivers: a review. **Biogeosciences**, v. 8, n. 4, p. 89-112, 2011.

COSTA, A. F. S.; ALVES, F. L.; COSTA, A. N. **Plantio, formação e manejo da cultura do maracujá.** In: COSTA, A. F. S.; COSTA, A. N. (Ed). Tecnologia para a produção do maracujá, p. 23- 53, 2005.

DIAS, T. J.; CAVALCANTE, L. F.; LEON, M. J.; FREIRE, J. L. O.; MESQUITA, F. O.; SANTOS, G. P.; ALBUQUERQUE, R. P. F. Produção do maracujazeiro e resistência mecânica do solo com biofertilizante sob irrigação com águas salinas. **Revista Ciência Agronômica**, v. 42, n. 3, p. 644-651, 2011.

DINIZ, K.A.; GUIMARÃES, S.T.M.R.; LUZ, J.M.Q. Humus as substrate for the production of tomato, pepper and lettuce seedlings. **Bioscience Journal**, v. 22, n. 3, p. 63-70, 2006.

FREIRE, J. L. O.; DIAS, T. J.; CAVALCANTI, L. F.; NETO, P. D.; LIMA, A. J. Quantum yield and gas exchange in yellow passion fruit tree under water salinity, biofertilization and mulching. **Revista Ciência Agronômica**, v. 45, n. 1, p. 82-91, 2014.

FERRAZ, R. L. S.; MAGALHÃES, I. D.; BELTRÃO, N. E. M.; MELO, A. S.; NETO, J. F. B.; ROCHA, M. S. Photosynthetic pigments, cell extrusion and relative leaf water content of the castor bean under silicon and salinity. **Revista Brasileira de Engenharia Agrícola e Ambiental**, v. 19, n. 9, p. 841-848, 2015.

FREITAS, L. B. Adubação foliar com silício na cultura do milho**. Revista Ceres**, v. 58, p. 262-267. 2011.

FALEIRO, F. G.; FARIAS NETO, A. L. E RIBEIRO JÚNIOR, W. Q. **Pré-melhoramento, melhoramento e pós melhoramento:** estratégias e desafios. 1st ed. Planaltina, Embrapa Cerrados, 2008, p. 183.

FREIRE, J. L. O.; DIAS, T. J.; CAVALCANTE, L. F.; FERNANDES, P. D.; LIMA NETO, A. J. Quantum yield and gas exchange in yellow passion fruit under water salinity, biofertilization and mulching. **Revista Ciência Agronômica**, v. 45, n. 1, p. 82-91, 2014.

GRECO, S. M. L. **Physical-chemical and molecular characterization of passion fruit genotypes cultivated in the Federal District**. 2014. 149f. Thesis - Faculty of Agronomy and Veterinary Medicine, Brasília, 2014.

GHEYI, H. R.; DIAS, N. S.; LACERDA, C. F. **Manejo da salinidade na agricultura:** Estudos básicos e aplicadas. Fortaleza: INCT Sal, 2010. 472p.

GONDIM, S.C.; LIMA, E.M.; MACEDO, J.P.S.; SANTOS, J.B.; SANTOS, C.J.O. **Maracujá-Amarelo e a Salinidade Some tropical fruits and salinity.** In: CAVALCANTE, L.F.; LIMA, E.M. (Org.). Jaboticabal: FUNEP, 2006. p. 91-115.

GONG, H.; ZHU, X.; CHEN, K.; WANG, S.; ZHANG, C. Silicon alleviates oxidative damage of wheat plants in pots under drought. **Plant Science**, v. 169, n. 2, p. 313-321, 2005.

HASHEMI, A.; ABDOLZADEH, A.; SADEGHIPOUR, H. R. Beneficial effects of silicon nutrition in alleviating salinity stress in hydroponically grown canola, Brassica napus L., plants. **Soil Science and Plant Nutrition**, v. 56, n. 2, p. 244-253, 2010.

HAGHIGHI, M.; PESSARAKLI, M. Influence of silicon and nano-silicon on salinity tolerance of cherrytomatoes (*Solanum lycopersicum* L.) at early growth stage. **Horticultural Science**, v. 161, n. p. 111-117, 2013.

HATTORI, T.; INANAGA, S.; ARAKI, H.; MORITA, S.; LUXOVÁ, M.; LUX, A. Application of silicon enhanced drought tolerance in *Sorghum bicolor*. **Physiologia Plantarum**, v. 123, n. 4, p. 459-466, 2005.

_____IBGE. Censo Agro, Paraíba, 2017a. Available at: < https://censos.ibge.gov.br/agro/2017/templates/censo_agro/resultadosagro/agricultura.html?localidade=25&tema=76346>. Accessed on : 8 Nov 2018.

_____IBGE. Censo Agro, Brazil, 2017b. Available at: < https://censos.ibge.gov.br/agro/2017/templates/censo_agro/resultadosagro/agricultura.html?localidade=0&tema=76346>. Accessed 8 Nov 2018.

JAYAKANNAN, M.; BOSE, J.; BABORINA, O.; SHABALA, S.; MASSART, A.; POSCHENRIEDER, C.; RENGEL, Z.The NPR1-dependent salicylic acid signalling pathway is pivotal for enhanced salt and oxidative stress tolerance in Arabidopsis. **Journal of Experimental Botany**, v. 66, n. 7, p. 1865- 1875, 2015.

LIANG, Y.; NIKOLIC, M.; BÉLANGER, R. Silicon in Agriculture. In: Silicon Uptake and Transport in Plants: Physiological and Molecular Aspects**. Springer Science,** p.69-82, 2015.

LEE, S.K; SOHN, E. Y; HAMAYUN, M.; YOON, J.Y.; LEE, I. J.; Effect of silicon on growth and salinity stress of soybean plant grown under hydroponic system. **Agroforestry Systems**, v. 80, n. 3, p. 333-340, 2010.

LENZA, J. B.; VALENTE, J. P.; RONCATTO, G.; ABREU, J. Al. Development of passion-fruit seedlings mudas de maracujazeiro propagadas por enxertia. **Revista Brasileira de Fruticultura,** v.31, n.4, p. 1135-1140, 2009.

MANAI, J.; KALAI, T.; GOUIA, H.; CORPAS, F. J. Exogenous nitric oxide (NO) ameliorates salinity-induced oxidative stress in tomato (*Solanum lycopersicum*) plants. **Journal of Soil Science and Plant Nutrition**, v.14, n. 2, p.433-446, 2014.

MOUSSA, H.R.; GALAD, M.A.R. Comparative Response of salt Tolerant and salt Sensitive Maize (*Zea mays* L.) Cultivars to Silicon. **European Jounal of Academic Essays**, v.2, n.1, p.1-5, 2015.

MEDEIROS, S.S.; CAVALCANTE, A.M.B.; MARIN, A.M.P.; TINÔCO, L.B.M.; SALCEDO, I.H. & Pinto, T.F. **Synopsis of the demographic sense for the Brazilian semiarid.** Campina Grande: INSA, 2012. 103p.

MUNNS, R.; TESTER, M. Mechanisms of salinity tolerance. **Annual Review of Plant Biology**, v.59, s.n., p.651-681, 2008.

OLIVEIRA, L.A. **Silicon in bean and rice plants: uptake, transport, redistribution and tolerance to cadmium**, USP, 2009. 157p. Thesis (Doctorate) - Center for nuclear energy in agriculture. University of São Paulo, Piracicaba, SP, Brazil.

OLIVEIRA, F. A.; LOPES, M. A. C.; SÁ, F. V. S.; NOBRE, R. G.; MOREIRA, R. C. L.; SILVA, L. A. S.; PAIVA, E. P. Interaction salinity of irrigation water and substrates in the production of yellow passion-fruit seedlings. **Revista Comunicata Scientiae**, v. 6, n. 4, p. 471-478, 2015.

PIRES, M.M.; MATA, H.T.C. **Uma abordagem econômica e mercadológica para a cultura do maracujá no Brasil.** In: Maracujá: produção e qualidade na passicultura. LIMA, A.A.; CUNHA, W. A.P. (editores). Cruz das Almas: Embrapa Mandioca e Fruticultura, 2004. p.323-343.

PIRES, A. A.; MONNERAT, P. H.; MARCIANO, C. R.; PINHO, L. G. R.; ZAMPIROLLI, P. D.; ROSA, R. C. C.; MUNIZ, R. A. Efeito da adubação alternativa do maracujazeiro amarelo nas características químicas e físicas do solo. **Revista Brasileira de Ciência do Solo**, v.32, p. 1997-2005, 2008.

QADOS, A. M. S. A.; MOFTAH, A. E. Influence of Silicon and Nano Silicon on Germination, Growth and Yield of Faba Bean (*Vicia faba* L.) Under Salt Stress Conditions. **American Journal of Experimental Agriculture**, v.6, n.5, p.509-524, 2015.

RASOOL, S.; HAMEED, A.; AZOOZ, M.M. et al. **Salt Stress: Causes, Types and Responses of Plants**. In: AHMAD, P. et al. (eds.), Ecophysiology and Responses of Plants under Salt Stress. 2013. p.1-24.

RIBEIRO, A. A.; FILHO, M. S.; MOREIRA, F. J. C.; MENEZES, A. S.; BARBOSA, M. C. Effect of salinity on the initial growth of yellow passion fruit (*Passiflora edulis* sims. f. flavicarpa deg.). **Revista Agrogeambiental**, v. 6, n. 3, 37-44, 2014.

ROHANIPOOR, A.; NOROUZI, M.; MOEZZZI, .; HASSIBI, P. Effect of Silicon on Some Physiological Properties of Maize (*Zea mays*) under Salt Stress**. Journal Biodiversity and Environmental Sciences**, v. 7, p.71-79, 2013.

RHOADES, J. D.; KANDIAH, A.; MASHALI, A. M. **The use of saline waters for crop production**. Rome: FAO, 1992. 133 p.

SÁ, F.V.S.; BRITO, M.E.B.; MELO, A.S.; ANTÔNIO NETO, P.; FERNANDES, P. D.; FERREIRA, I.B. Produção de mudas de mamoeiro irrigadas com água salina. **Revista Brasileira Engenharia Agrícola e Ambiental**, v. 17, n. 10, p. 1047-1054, 2013.

SALATA, A.C.; HIGUTI, A.R.O.; GODOY, A.R.; MAGRO, F.O.; CARDOSO, A.I.I. Zucchini production as a function of seedling age. **Ciência e agrotecnologia**, v. 35, n. 3, p. 511-515, 2011.

SILVA, R.M. **Produção de mudas de maracujazeiro-amarelo com diferentes tipos de enxertia e uso de câmera úmida**, UFERSA, 2012. 59p. Dissertation (Master's Degree) - Universidade Federal Rural do Semiárido, Mossoró, RN.

ROHANIPOOR, A.; NOROUZI, M.; MOEZZZI,.; HASSIBI, P. Effect of Silicon on Some Physiological Properties of Maize (*Zea mays*) under Salt Stress**. Journal Biodiversity and Environmental Sciences**, v. 7, p.71-79, 2013.

RIBEIRO, A. A.; FILHO, M. S.; MOREIRA, F. J. C.; MENEZES, A. S.; BARBOSA, M. C. Effect of salinity on the initial growth of yellow passion fruit (*Passiflora edulis* sims. f. flavicarpa deg.). **Revista Agrogeambiental**, v. 6, n. 3, 37-44, 2014.

SANGSTER, A.G.; HODSON, M.J.; PARRY, D.W. Silicon deposition and anatomical studies in the inflorescence bracts of four Phalaris species with their possible relevance to carcinogenisis. **New Phytologist**, v.93, n. 1, p.105-122, 2001.

SAIRAM, R. K.; TYAGY, A. Physiology and molecular biology of salinity stress tolerance in plants. **Current Science**, v. 86, n. 3, p. 407-421, 2004.

STEPPUHN, H. Pre-irrigation of a severely-saline soil with in situ water to establish dry land forages. **Transactions of the ASAE**, v. 44, n. 6, p. 1543- 1551, 2001.

SIVANESAN, I.; SON, M. S.; LIM, C. S.; JEONG, B. R. Effect of soaking of seeds in potassium silicate and uniconazole on germination and seedling growth of tomato cultivars, Seogeon and Seokwang. **African Journal of Biotechnology**, v. 10, n. 35, p. 6743-6749, 2011.

SOUSA, V. F.; FOLEGATTI, M. V.; COELHO, M. A.; FRIZZONE, J. A. Distribuição radicular do maracujazeiro sob diferentes doses de potássio aplicado por fertirrigação. **Revista Brasileira de Engenharia Agrícola e Ambiental**, v. 6, n.1, p. 51-56, 2002.

SOUSA, G. B. et al. Initial growth of the yellow passion fruit tree irrigated with saline water in substrate with bovine biofertilizer. In*:* **CONGRESSO BRASILEIRO DE FRUTICULTURA**, 19, 2006, Cabo Frio. Lecture and Abstracts. Cabo Frio: SBF, UENF, UFRuralRJ ,2006. p. 411.

SHI, Y.; WANG, Y.; FLOWERS, T.J.; GONG, H. Silicon decreases chloride transport in rice (*Oryza sativa* L.) in saline conditions. **Journal of Plant Physiology**, v.170, p. 847-853, 2013.

TAIZ, L.; ZEIGER, E. **Fisiologia vegetal**. 5. ed. Porto Alegre: Artmed, 2013. 918 p.

TAHIR, M. A.; AZIZ, T.; FAROOQ, M.; SARWAR, G. Silicon-induced changes in growth, ionic composition, water relations, chlorophyll contents and membrane permeability in two salt-stressed wheat genotypes**. Archives Agronomy Soil Science**, v. 58, n. 3, p. 247-56, 2012.

YIN, L. Application of silicon improves salt tolerance through ameliorating osmotic and ionic stresses in the seedling of Sorghum bicolour. **Acta Physiol Plant**, v. 35, n. 11, p. 1-9, 2013.

YADAV, K.S.; NASEERUDDIN, S.; PRASHANTHI, G.S.; SATEESH, L.; RAO, L.V. Bioethanol fermentation of concentrated rice straw hydrolysate using co-culture of

Saccharomyces cerevisiae and Pichia stipitis. **Bioresourse Technology**, v. 102, n. 11, p. 6473-6478, 2011.

VIANA, P. C.; LIMA, J. G. A.; ALVINO, F. C. G.; JUNIOR, J. R. S.; GOMES, E. C.; VIANA, K. C. Efeito da salinidade da água de irrigação na produção de maracujazeiro-amarelo. **Agropecuária científica no semárido**. v. 8, n. 1, p. 45-50, 2012.

CHAPTER I

OSMOTIC POTENTIAL AND PHYSIOLOGICAL INDICES OF PASSION-FRUIT SEEDLINGS UNDER SALT STRESS AND SILICATE FERTILIZATION

Resumo: O Nordeste Brasileiro sofre com alguns estresses abióticos que são responsáveis pela perda de produção agrícola, especialmente na parte semiárida, que apresentam longos períodos de estiagem e elevada evapotranspiriração o que induzca ao uso de águas salinas como alternativa para expansão das áreas irrigadas, com isto da adubação silicatada exerce papel importante na atenuação do estresse salino para as condições de semárido Nordestino. Neste sentido, objetivou-se com este trabalho avaliar o potencial osmótico e os índices fisiológicos do maracujazeiro amarelo em função da salinidade da água de irrigação e da adubação silicatada. The experiment was developed in vegetation house conditions in the Center of Science and Technology Agroalimentary, in the city of Pombal-PB. The cultivar studied was the yellow giant passion-fruit tree. A randomized block design was used in a 5 x 5 factorial scheme, with five levels of electrical conductivity of the irrigation water (0.3; 1.0; 1.7; 2.4 and 3.1 dS m-1) and five doses of silicate fertilization (0; 25; 50; 75 and 100 of potassium silicate plant-1), with four repetitions and two plants per plot, totalizing two hundred experimental units. The photosynthetic pigments (Chlorophyll *a, b* and carotenoids), photochemical efficiency (initial fluorescence, maximum fluorescence, variable and quantum efficiency of photosystem II) and leaf osmotic potential were analysed. The data obtained were subjected to the F test at 0.01% and 0.05% probability, when there was significant effect of treatments, the means were subjected to polynomial regression for the saline levels and for the doses of silicon. The salinity in irrigation water reduced the photosynthetic pigments and yield in photosystem II in yellow passion-fruit plants. The silicate fertilization at a dose of 75 g plant-1 attenuated the deleterious effect of salt stress on osmotic potential, chlorophyll b and carotenoids.

Key words: *Passiflora edulis* f. flavicarpa, salinity, silicon.

Abstract: Brazilian Northeast suffers from some abiotic stresses that are responsible for the loss of agricultural production, especially in the semiarid areas, which are exposed to long periods of drought and high evapotranspiration, which induces the use of saline water as an alternative for the expansion of irrigated areas. Silicate fertilization plays an important role in attenuating saline stress for Northeastern semiarid conditions. In this sense, this work aims to evaluate the osmotic potential and the physiological indices of yellow passion fruit in function of the salinity of the irrigation water and the silicate fertilization. The experiment was carried out under greenhouse conditions at the Center for Science and Agri-Food Technology, in the municipality of Pombal-PB. The studied cultivar was the giant yellow passion fruit. The experimental design was a randomized blocks 5 x 5 factorial scheme, whose treatments consisted of five levels of electrical conductivity of the irrigation water (0.3; 1.0; 1.7; 2.4 and 3, 1 dS m-1) and five doses of silicate fertilizer (0; 25; 50; 75 and 100 of potassium silicate plant-1) with four replications and two plants per plot, totalizing two hundred experimental units. Photosynthetic pigments (Chlorophyll a, b and carotenoids), photochemical efficiency (initial fluorescence, maximum fluorescence, variable and quantum efficiency of photosystem II) and leaf osmotic potential were analyzed. The obtained data were submitted to the F test at 0.01% and 0.05% probability. When there was a significant effect of the treatments, the averages were submitted to polynomial regression for the saline levels and for the doses of silicon. Salinity in irrigation water reduced photosynthetic pigments and yield in photosystem II in yellow passion fruit plants. Silicate fertilization at a 75 g plant-1 dose attenuated the deleterious effect of salt stress on osmotic potential, chlorophyll b and carotenoids.

Key words: Passiflora edulis f. flavicarpa, salinity, silicon.

1 INTRODUCTION

The passion-fruit (*Passiflora* spp.) is cultivated in almost all Brazilian territory, by small, medium and large producers and mainly in family farming, generating employment and income for farmers, having great socioeconomic expression by having domestic and foreign consumer market, with various products in the form of juices, jams and jellies (MELETTI, 2011).

The yellow passion-fruit is considered a short-cycle plant, with production beginning between 6 and 9 months after planting (SILVA, 2012). Thus, this crop demands a greater investment from the producer in the acquisition of high quality seedlings. According to Leitão et al. (2009) the production of seedlings is a crucial phase in the development of the plants so that good yields can be obtained in the Northeast region, where the availability and quality of water are limiting factors for agricultural cultivation.

Moreover, the Northeast region has low rainfall and high losses of water by evaporation, presenting more than 60% of arid climate areas (MEDEIROS et al., 2012), which leads to high concentrations of salts in groundwater, artesian wells, rivers and dams, affecting the quality of water that is intended for irrigation, among other purposes.

In situations of salt stress, the osmotic and ionic effects influence plant development, causing alterations in the metabolic activities of the cells and in the cell elongation process, compromising plant growth and may even lead to death (SAÍRAM AND TYAGI, 2004; TAIZ AND ZEIGER, 2013).

In order to mitigate the effects of salinity on the development of crops, some researches have been carried out using silicon (Si) as a salt stress attenuator. Silicon has several benefits for crops in tolerating salt stress, such as increasing the photosynthetic rate and the capacity of antioxidant defense against water deficiency (HATTORI et al., 2005). Studies have shown beneficial effects of Si in several crops (soybean, tomato, corn) when subjected to salt stress conditions, where an increase in the leaf area, chlorophyll content and improvement of the chloroplast structure was observed, which provided an increase in photosynthetic activity (HASHEMI et al., 2010; LEE et al., 2010; TAHIR et al., 2012).

In this context, the objective was to evaluate the osmotic potential, the photosynthetic pigments and the photochemical efficiency of yellow passion-fruit as a function of irrigation with water of different salinities and silicate fertilization under conditions of the semi-arid region of Paraiba.

2 MATERIAL AND METHODS

2.1 Site of the experiment

The experiment was conducted under greenhouse conditions at the premises of the Centre for Agro-Food Sciences and Technology of the Federal University of Campina Grande, in the municipality of Pombal-PB, located at 6°47'3" S, 37°49'15" W and 193 m altitude.

2.2 Experimental procedure

A randomized block design was adopted, with treatments arranged in a 5 x 5 factorial scheme relating to five levels of electrical conductivity of the irrigation water - CEa (0.3; 1.0; 1.7; 2.4 and 3.1 dS m-1) and five doses of silicate fertilization (0; 25; 50; 75 and 100 g of potassium silicate per plant) with two plants per plot and four repetitions, totaling 200 experimental units. The salinity levels of the water were determined based on research developed by Andrade (2018). The levels of ECa were obtained from the addition of sodium chloride (NaCl) in the water supply of the city of Pombal-PB (0.3 dS m-1) obeying the relationship between ECa and the concentration of salts (mg L-1 = 640 x ECa) (RICHARDS, 1954).

Potassium silicate was used for the fertilizer doses. It is a compound of multi-minerals: selenium, vanadium, calcium, zinc, phosphorus and various trace elements, with 50% SiO_2 and 4% K_2O.

The culture which was studied was the Yellow Giant passion fruit, a hybrid adapted to altitudes ranging from 376 to 1100m, which can be planted at any time of year. Destined to the fruit market and industry. It has bright yellow fruits, weighing between 120 and 350 grams. The pulp represents 40% of the fruit, with a strong yellow colour and a high percentage of vitamin C. °Brix between 13 and 15 degrees.

The research was initiated with sowing in plastic bags with dimensions of 15 x 20 cm with capacity of 1250 ml, filled with substrate 2:1:1 in volume base (soil, sand and tanned bovine manure). The soil moisture content was raised to field capacity with water of the lowest saline level (0.3 dS m-1), where water was placed in the bags until free drainage occurred. Two seeds were sown per bag and after seedling emergence, only one plant per container was thinned when they were 10 cm high.

Irrigation was performed daily manually, applying water from the respective treatment and based on the process of drainage lysimetry (BERNARDO et al., 2006). The volume applied in each irrigation was determined by the difference between the volume applied and the drained volume of the previous day, plus a leaching fraction of 15% applied every 20 days. The application of saline levels began at 30 DAS and was performed daily until 60 DAS.

The silicon doses were applied diluted via irrigation water and started at 30 DAS and were carried out weekly, totaling 4 applications until 60 DAS, according to the treatments, where 0; 6.25; 12.5; 18.75 and 25g were applied weekly per plant.

The physical and chemical characteristics of the soil used for the production of seedlings were analysed by the laboratory of irrigation and salinity of UFCG, as shown in the results of (Table 1).

Table 1: Physical and chemical characteristics of the soil used for the production of yellow passion-fruit seedlings, carried out by the laboratory of irrigation and salinity of the UFCG - Campina Grande-PB campus. Pombal-PB.

		.... Chemical attributes							
Ph		P		K+	Na+	Ca2+	Mg2+	Al3+	H + Al3+
CaCl2 1:2.5	ECes (dS m-1)	mg dm-3		............cmolc dm-3					
7,00	0,20	0,21		0,38	0,09	2,50	3,75	0	0
....		.... Physical Attributes							
Sand		Silt	Clay	Ds	Dp	Porosity	UD		
	............g kg-1			..... kg dm-3.....		%	%	Textural class	
85,30		13,07	1,63	1,50	2,69	47,23	0,55	Areia Franca	

pHes = pH of the saturation extract of the substrate; ECes = Electrical conductivity of the saturation extract of the substrate at 25 °C. Ca2+ and Mg2+ extracted with 1 M KCl pH 7.0; Na+and K+ extracted using 1 M NH4OAc pH 7.0; H+ and Al3+ extracted with 0.5 M CaoAc pH 7.0; FA - Clayy loam; AD - Available water. Ds - Soil density, Dp - Particle density. UD - Moisture (dry soil base).

Fertilization with nitrogen and potassium was carried out weekly 20 DAS, for the fertilization with potassium the content of the element already present in the potassium silicate was taken into consideration. The fertilization with phosphorus was applied in the foundation according to Santos (2001). Where urea was used as a source of nitrogen, MAP as a source of phosphorus and potassium chloride as a source of potassium.

Fertilization description: NPK fertilization was carried out using 100 mg N kg-1 soil, 150 mg K2O kg-1 soil and 300 mg P2O5 kg-1 soil. The following fertilizers were applied per plant: 0.217 g of urea, 1.4 g of monoammonium phosphate and 0.56 g of potassium chloride, divided into 4 portions, supplied via fertigation at 20 days after planting.

The micronutrients were applied weekly at 30 DAS using the product Quimifol (water, nitrogen solution, boric acid, lignosulfonates, amino acids, polysaccharides, leonardite and preservative) applying 0.5 g per litre (Cavalcanti, 2008). During the experiment, the cultural and phytosanitary treatments recommended for the crop, such as pest control and removal of tendrils, were carried out.

2.3 Variables analysed

The effects of the treatments on the passion-fruit seedlings were analysed at 60 days after sowing (DAS), a period in which the seedlings were ready to be sent to the field. The foliar osmotic potential, photosynthetic pigments (chlorophyll *a, b* and carotenoids) and photochemical efficiency (initial fluorescence - Fo, maximum - Fm, variable - Fv and photosystem II quantum efficiency - Fv/Fm) were evaluated.

To determine the osmotic potential of the leaf sap, the leaf sap was extracted by placing leaves into an Eppendorf tube previously perforated at the base, which functioned as a minifilter. With a glass rod, the leaf tissue was pressed, resulting in the extraction of the sap that was collected in another Eppendorf tube and then the extract was centrifuged at 10,000 g for 10 min at 4 ºC. A 10 μL aliquot of the supernatant was used to determine the osmolality of the passion fruit leaf tissue using a vapor pressure osmometer, model Wescor® 5520. The osmotic potential values were obtained from the osmolality (mmol kg-1) of the sap of the leaf tissue, using the Van't Hoff equation (SOUZA et al., 2012) and then converted into Mpa, according to eq. 1:

$$\psi s\ (\text{MPa}) = -C\left(\frac{mOsmol}{kg}\right) x\ 2{,}58\ x\ 10^{-3} \qquad (1)$$

where:

ψs (MPa) = leaf osmotic potential

C= osmolality of the sample, found in the osmometer reading.

The chlorophyll *a* and *b* and carotenoids contents were determined according to the methodology developed by Arnon (1949), by means of samples of 5 discs of the limb of the third mature leaf from the apex. From the extracts, the concentrations of chlorophyll and carotenoids in the solutions were determined by means of spectrophotometer at absorbance wavelength (ABS) (470, 646, and 663 nm), using the equations:

Chlorophyll *a* (Cl *a*) = 12.21 ABS663 - 2.81 ABS646;

Chlorophyll *b* (Cl *b*) = 20.13 A646 - 5.03 ABS663;

Total carotenoids (Car) = (1000 ABS470 - 1.82 Cl *a* - 85.02 Cl *b*) /198.

The values obtained for chlorophyll *a*, *b* and carotenoids contents in the leaves were expressed in mg g-1 of fresh matter (mg g-1 MF).

The photochemical efficiency of passion-fruit was determined by initial fluorescence (Fo), maximum (Fm), variable (Fv), FSII quantum efficiency (Fv/Fm), in leaves pre-adapted to the dark using leaf clamps for 30 minutes, between 7:00 and 8:00 am, in the median leaf of the intermediate productive branch of the plant, using Opti Science model OS5p modulated fluorometer.

2.4 Data analysis

The data obtained in this research were submitted to the F test at 0.01 and 0.05% probability, to carry out variance analysis. When there was a significant effect of the treatments, the means of the variables were submitted to polynomial regression for the doses of silicon and salt levels. The statistical analyses were performed using the software SISVAR Version 5.6 (FERREIRA, 2011).

3 RESULTS AND DISCUSSION

There is significant interaction between the factors salt levels and doses of silicon (NS x DS) on the ψs and photosynthetic pigments (Cl *a*, Cl *b* and CAR), demonstrating that both factors act together on these variables. Rezende et al. (2017) when studying the interaction between doses of silicon and salt stress in plants of *Physalis peruviana*, also found an interactive effect on photosynthetic pigments (Table 2).

Table 2: Summary of the analysis of variance for the contents of chlorophyll *a* (CL *a*), *b* (CL *b)* and carotenoids (CAR) and osmotic potential (ψs), of passion fruit plants 'Gigante amarelo' grown under different levels of salinity of irrigation water and doses of silicon, at 60 days after sowing. Pombal, CCTA/UFCG, 2020.

Source of variation	GL	Mean Squares			
		CL *a*	CL *b*	CAR	Ψs
Salt levels (NS)	4	10,90*	1,05*	0,18ns	0,19**
Linear Reg.	1	16,59*	1.52ns	0,18ns	0,71**
Quadratic Reg.	1	0.44ns	1,25ns	0,21ns	0.005ns
Silicon dose (DS)	4	1,20ns	1,45*	0,64*	0.022ns
Linear Reg.	1	9,05ns	0.0034ns	0.095ns	0.022ns
Quadratic Reg.	1	0.49ns	1.07ns	0.09ns	0.046ns
Interaction (NSxDS)	16	8,47*	1,68**	0,44*	0,026*
Blocks	3	10,19*	1,80**	0.12ns	0.01ns
Waste	72	4,30	0,41	0,24	0,01
CV (%)		19,25	16,23	20,61	-11,10

ns, **, *, respectively not significant, significant at $p < 0.01$ and $p < 0.05$.

As for the osmotic potential (Figure 1A) a linear decreasing behaviour was observed with a reduction of 26.7%, 27.7% and 22.4% per unit increase in the electrical conductivity of the irrigation water in plants fertilized with 0, 25 and 75 g of silicon plant-1, respectively. The regression equation (Figure 1A) shows that the increase in the salinity of the water reduced the osmotic potential of the leaf sap, with the minimum estimated values of -1.07

and -1.16 Mpa in plants irrigated with conductivity water of 1.7 and 1.8 dS m-1 and fertilized with 50 and 100 g of silicon plant-1, respectively.

The reduction of the osmotic potential due to the increase of salts in the irrigation water is a response of the plant, in order to favor the potential gradient and contribute to the absorption of water and nutrients (CRUZ et al., 2018). It was observed that the plants fertilized with silicon at doses of 50, 75 and 100 g plant-1 had less reduction of osmotic potential, due to the deposition of silicon on the leaf wall increasing the resistance and hardness of the cell walls, reducing cuticular transpiration and consequently increasing the efficiency of water use, which is a positive point for acclimatization of the plant in saline conditions (JESUS et al., 2018).

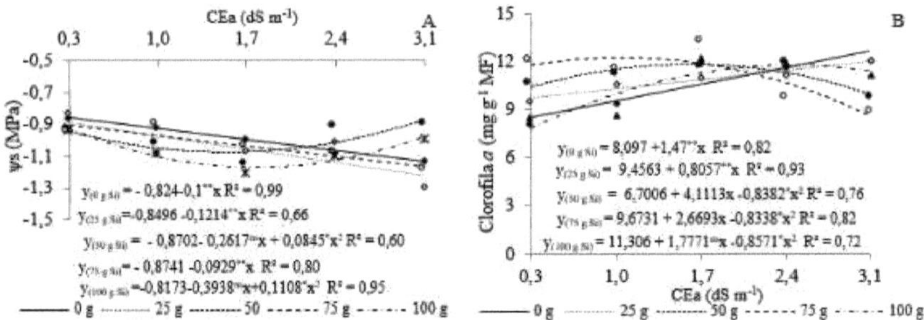

Figure 1. Osmotic potential of leaf sap - ψs (A) and chlorophyll content *a* (B) of passion-fruit plants 'Gigante amarelo' as a function of the interaction between water salinity levels - CEa and doses of silicate fertilizer, at 60 days after sowing. UFCG/CCTA, Pombal - PB, 2020.

The chlorophyll *a* (Figure 1B) of the yellow passion-fruit plants showed an increasing linear behavior, with an increase of 18.15% and 8.52% by unitary increase of the ECa, for plants submitted to doses of silicon of 0 and 25 g plant-1, respectively. The other doses (50, 75 and 100 g silicon) showed quadratic behavior with maximum estimated of

11.80, 12.22 and 11.74 mg g-1, in saline levels 1.6, 1.0 and 2.5 dS m-1, respectively, with a subsequent tendency to decrease.

Salt stress causes a reduction in osmotic potential, reducing the absorption of water and nutrients, which favors the closure of stomata, in order to mitigate water loss by transpiration, thus reducing photosynthetic activity (REIS et al., 2016). According to Ávila et al. (2010), Ferraz et al. (2014) and Costa et al. (2018) silicon mitigates these effects and by accumulating in the cell walls of plants reduces water loss through stomata and increases the photosynthetic rate, as well as the content of photosynthetic pigments.

The chlorophyll *b* contents (Figure 2A) of passion-fruit plants were also significantly affected by the interaction between the factors (NS x SD) and through the regression equations it was observed that the data had a better adjustment to the quadratic model, with the maximum values estimated for the chlorophyll *b* contents (4.41; 4.18; 4.53; 4.58 and 3.96 mg g-1) with a tendency to decrease from ECa levels of 0.7; 1.7; 0.8 and 1.6 dS m-1, in silicon doses of 25, 50, 75 and 100 g of silicon plant-1, respectively, demonstrating that the excess of salts can affect the content of chlorophyll *b in* passion-fruit leaves regardless of the concentration of silicon. However, the application of silicon in doses of 50 and 75 g in lower saline concentrations (1.7 dS m-1) favored greater increments in the content of chlorophyll *b*. According to the data, we can see that the doses of 50 and 75 g provided adequate levels of chlorophyll using water with ECa of 1.7 dS m-1 on average.

According to Munns and Tester (2008) the reduction in chlorophyll content is related to the deleterious effects caused by excess salts in irrigation water, due to imbalances in physiological and biochemical activities and stimulation of the activity of the chlorophyllase enzyme, which degrades the photosynthesizing pigment molecules, inducing the destruction of chloroplasts and favouring an unbalance as well as loss of activity of the pigmentation proteins. Cavalcante et al. (2011) when evaluating the contents of chlorophylls and carotenoids in yellow passion-fruit trees irrigated with saline water (0.5; 1.5; 2.5; 3.5 and 4.5 dS m-1) found that the increase in the salinity of irrigation water up to 2.5 dS m-1 did not compromise the biosynthesis of these pigments, however, levels of ECa greater than 2.5 dS m-1 decreased the photosynthetic efficiency of the plants.

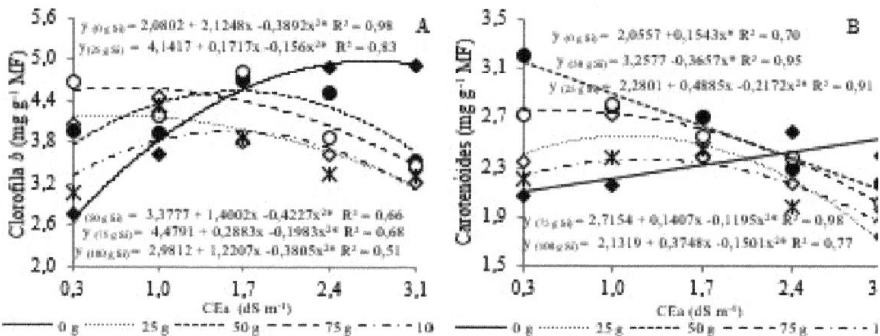

Figure 2: Chlorophyll *b* (A) and carotenoids (B) contents in passion-fruit plants 'Gigante Amarelo' as a function of interaction between water salinity levels - CEa and doses of silicon, 60 days after sowing. UFCG/CCTA, Pombal - PB, 2020.

In relation to the contents of carotenoids (Figure 2B), the regression equations showed that the plants fertilized with doses of 0 and 50 g of Si plant-1 fitted the increasing and decreasing linear model, respectively. When the passion-fruit plants did not receive doses of silicon (0 g plant-1) there was an increase in the carotenoids contents of 7.50% per unitary increase of ECa. Plants that received 50 g of Si plant-1 decreased carotenoids by 11.22% per unit increase in water salinity. On the other hand, the plants fertilized with 25, 75 and 100 g of Si plant-1 showed a quadratic behavior with a tendency to gradually increase the synthesis of carotenoids as a function of increasing salinity, reaching the maximum estimated values of 2.55; 2.75 and 2.36 mg g-1, in plants submitted to water salinity of 1.2, 0.8 and 1.4 dS m-1.

The increase in the contents of carotenoids in passion-fruit plants fertilized with doses of 25, 75 and 100 g of Si may be related to the fact that this element increases the antioxidant action from non-enzymatic compounds like carotenoids, significantly reducing the degradation of chlorophyll as well as the oxidative stress that is a consequence of excess salts (KIM et al., 2017). Divergent results were found by Wanderley et al. (2018) in research with passion-fruit culture under field or greenhouse conditions irrigated with saline water (ECa ranging from 0.3 to 3.1 dS m-1) where they observed that the increase in water salinity did not significantly influence the contents of carotenoids.

According to the summary of the analysis of variance for the variables initial flowering, variable, maximum and quantum efficiency of photosystem II (Table 3). There was an interactive effect of the factors saline levels of water and doses of silicon on all the variables analyzed, at 60 days after sowing, demonstrating that both factors act jointly on these variables.

Table 3. Summary of the analysis of variance for the initial fluorescence (Fo), variable (Fv), maximum (Fm) and quantum efficiency of photosystem II (Fv/Fm) of passion-fruit 'Gigante amarelo' cultivated under different levels of salinity of the irrigation water and doses of silicon, at 60 days after sowing. UFCG/CCTA, Pombal - PB, 2020.

Source of variation	GL	Mean Square			
		Fo	Fv	Fm	Fv/Fm
Salt levels (NS)	4	115798,46**	1322463,39**	541627,30**	0,036**
Linear Reg.	1	893.23ns	452390,71**	7248,08ns	0,11**
Quadratic Reg.	1	37955,71*	1082611,83**	132428,50ns	0.021ns
Silicon dose (DS)	4	12445,84ns	417603,48**	53218.72ns	0,028**
Linear Reg.		121734,44**	3871328,40**	1139287,50**	0,13**
Quadratic Reg.		5414.88ns	88007,14ns	38622,50ns	0.0054ns
Interaction (NSxDS)	16	33716,82**	342333,67**	112588,21**	0,037**
Blocks	3	4878.83ns	22033,46ns	33810.92ns	0.0053ns
Waste	72	6566,30	51111,19	42206,78	0,0061
CV (%)		9,42	8,78	5,77	10,83

ns, **, *, respectively not significant, significant at $p < 0.01$ and $p < 0.05$.

For the initial fluorescence of passion-fruit (Figure 3A) the regression equations showed that the plants that received doses of silicon of 0, 25 and 50 g plant-1 obtained maximum estimated values of 998.24, 942 and 955.7 when irrigated with conductivity water of 1.1, 0.8 and 0.3 dS m-1, with reduction from these levels of CEa. On the other hand, the passion-fruit plants that received doses of 75 and 100 g of silicon plant-1 showed a linear decreasing behavior with a reduction of 2.76% and 7.82% per unitary increment of ECa.

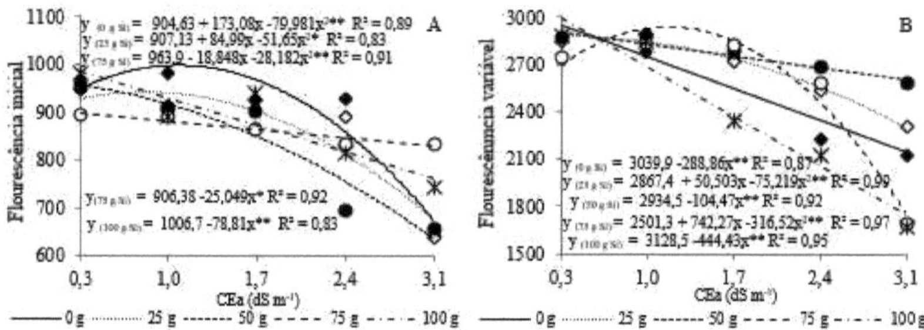

Figure 3 - Initial fluorescence - Fo (A) and variable (B) of passion-fruit plants 'Gigante Amarelo' as a function of the interaction between the saline levels of water - CEa and doses of silicon, at 60 days after sowing. UFCG/CCTA, Pombal - PB, 2020.

The reduction in fluorescence in plants grown under salt stress conditions is caused by changes in the state of the thylakoid membranes of chloroplasts, thus triggering changes in the characteristics of fluorescence signals (LIMA et al., 2018). Probably, silicon concentrations higher than 25g provided greater reductions in carbon uptake with increasing salinity of irrigation water, entailing mitigation in the photochemical capacity of passion fruit (TAIZ et al., 2017). Perhaps due to an oxidative stress.

According to Lucena et al. (2012) the initial fluorescence has an intrinsic relationship with fluorescence when the quinone, which is the primary electron receptor of photosystem II is completely oxidized, while the centre of reactions is open, thus indicating the activation of photochemical reactions in the cells. Thus, saline levels above 1.1 dS m-1 affect the performance of photosystem II in passion-fruit plants, further indicating the presence of this stress.

Regarding the variable fluorescence of passion-fruit (Figure 3B) it was observed that the plants fertilized with doses of 0, 50 and 100 g of Si plant-1 fitted the linear regression model, whose decreases were of 27.38%, 10.07% and 41.54%, respectively, per unit increase of ECa. The plants submitted to fertilization with 25 and 75 g of Si plant-1 showed a quadratic behavior, with maximum estimated values of 2871 and 2936 obtained in plants under ECa of 0.6 and 1.2 dS m-1, respectively.

Regarding maximum fluorescence (Figure 4A) a linear decreasing tendency was observed for the plants that received the doses of 0, 25, 50 and 100 g Si plant-1, whose reduction was of 3.45%; 2.5%; 2.8% and 4.35% per unit increase in ECa, equivalent to a reduction of 13.8%; 10.0%; 11.2% and 17.4%, respectively, among those under irrigation with water of conductivity of 0.3 and 3.1 dS m-1. The plants fertilized with 75 g Si plant-1 showed quadratic behavior, having obtained the maximum estimated value of 3682.2 when they received water salinity of 1.3 dS m-1.

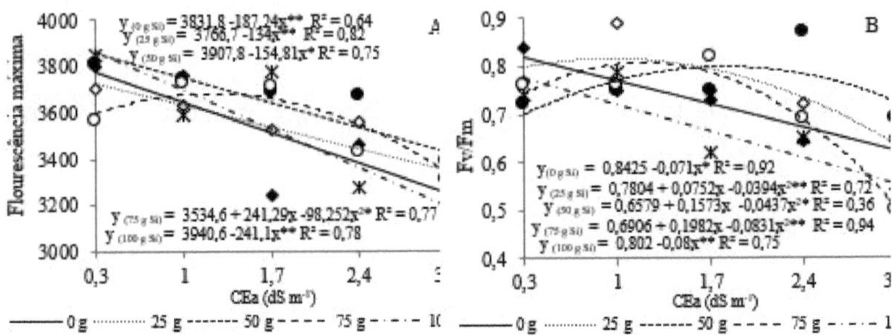

Figure 4. maximum fluorescence (A) and quantum efficiency of photosystem II (Fv/Fm) (B) of passion-fruit 'Gigante Amarelo' plants as a function of the interaction between saline levels of water - CEa and doses of silicon, at 60 days after sowing. UFCG/CCTA, Pombal - PB, 2020.

Similarly to what was observed for variable fluorescence (Figure 3B), the dose of 75 g of Si plant-1 also provided higher Fm in passion-fruit cultivated with saline water of up to 1.3 dS m-1. According to Kim et al. (2017) the application of silicon favours the elimination of ROS and as a consequence activates the components responsible for the plant defence system, which are the enzymes catalase and ascorbate peroxidase, involved in the conversion of hydrogen peroxide into water. This application also induces the second form of defence used by plants, which is through carotenoids.

The quantum efficiency of photosystem II in passion-fruit plants (Figure 4B) was linear for plants submitted to doses of 0 and 100 g of Si plant-1, with a reduction of 24.2% and 28.7%, respectively, between the highest and lowest saline levels studied. However, the doses of 25, 50 and 75 g of silicon showed quadratic behavior with maximum estimated values of 0.81, 0.79 and 0.80 in plants that received water conductivity of 1.3, 1.8 and 1.5 dS m-1.

The reduction in the quantum efficiency of photosystem II was observed in passion-fruit plants at all doses up to the maximum of silicon (100 g) as a function of different salt levels may be explained by the consequences of Na- and Cl- accumulation in the chloroplasts, which directly affects the photosynthetic process, causing alterations in the total content of pigments with chlorophyll, reducing the performance of photosynthetic enzymes and also limits the transport of electrons in chloroplasts, reflecting in decreases in the photochemical efficiency of photosystem II (HUANG et al., 2012).

Freire et al. (2014) evaluating the quantum yield and gas exchange in yellow passion-fruit under water salinity, biofertilizer and mulching, verified that salt stress inhibited the photosynthetic activity of photosystem II and net photosynthesis of the plants. Also according to these authors, salinity promotes damage to the photosynthetic apparatus of passion-fruit plants, compromising photosystem II as a function of the time of exposure of the plants to abiotic stress.

4 CONCLUSIONS

Plants fertilized with silicon at doses of 50, 75 and 100 g plant-1 showed less reduction in osmotic potential; application of silicon at doses of 50 and 75 g.

Saline concentrations lower than 1.3 dS m-1 favoured greater increases in chlorophyll b content; an increase in carotenoids was observed in plants fertilized with 25 and 100 g of silicon;

Salt levels above 1.1 dS m-1 affected the photosystem II performance of passion-fruit plants when subjected to doses of silicon;

5 BIBLIOGRAPHICAL REFERENCES

ANDRADE, J.R.; SOUSA, A.; JUNIOR, M.; OLIVEIRA, S.; PAULA; REZENDE, L. P.; ARAÚJO NETO, J. C. Germination and morphophysiology of passion fruit seedlings under salt water irrigation. **Tropical Agricultural Research**, v. 48, n. 3, p. 229-236, 2018.

ARNON, D. I. Copper enzymes in isolated chloroplasts: polyphenoloxidases in *Beta vulgaris*. **Plant Physiology**, v.24, n.1, p.1-15, 1949.

ÁVILA, F.W.; BALIZA, D.P.; FAQUIN, V.; ARAÚJO, J.L.; RAMOS, S.J. Interaction between silicon and nitrogen in rice grown under nutritive solution. **Revista Ciência Agronômica**, v.41, p.184-190, 2010.

BERNARDO, S; SOARES, A. A.; MANTOVANI, E. C. Manual de irrigação. Viçosa: UFV, 2006. 625p.

CAVALCANTE, L. F.; DIAS, T. J.; NASCIMENTO, R. FREIRE, J. L. O. Chlorophyll and Carotenoids in Yellow Passionflower Irrigated with Saline Water in Soil with Bovine Biofertilizer. **Revista Brasileira de Fruticultura**, volume especial, p.699-705, 2011.

CAVALCANTI, J. C. P. **Recomendações de adubação para o estado Pernambuco (2ªaproximação)**. 3.ed. Recife: Instituto agronômico do Pernambuco - IPA, 2008. 212p.

COSTA, B. N. S.; COSTA, I. J. S.; DIAS, G. M. G.; ASSIS, F. A.; PIO, L. A. S.; SOARES, J. D. R.; PASQUAL, M. Morphoanatomical and Physiological Modifications of Passionflower fertilized with silicon. **Revista Agropecuária Brasileira**, v.53, n. 2, p.163-171, 2018.

CROCHEMORE, M. L.; MOLINARI, H. B.; STENZEL, N. M. C. Caracterizacao agromorfológica do maracujazeiro (*Passiflora* spp.). **Revista Brasileira de Fruticultura,** v. 25, n. 1, p. 5-10, 2003.

CRUZ, A.F.S.; SILVA, G.F.; SILVA, E.F.F.; SOARES, H.R.; SANTOS, J.S.G.; LIRA, R.M. Stress index, water potentials and leaf succulence in cauliflower cultivated hydroponically with brackish water. **Brazilian Journal of Agricultural and Environmental Engineering**, v.22, n.9, p.622-627, 2018.

FERRAZ, R.L. de S.; BELTRÃO, N.E. de M.; MELO, A. S. de; MAGALHÃES, I.D.; FERNANDES, P.D.; ROCHA, M. do S. Gas exchange and photochemical efficiency of herbaceous cotton cultivars under foliar silicon application. **Semina: Agricultural Sciences**, v.35, n. 2, p.735-748, 2014.

FERRAZ, R. L. S.; MAGALHÃES, I. D.; BELTRÃO, N. E. M.; MELO, A. S.; BRITO N. J. F.; Rocha, M. S. Photosynthetic pigments, cell extrusion and relative leaf water content of the castor bean under silicon and salinity. **Revista Brasileira de Engenharia Agrícola e Ambiental**, v.19, n.9, p.841-848, 2015.

FERREIRA, D. F. SISVAR: A computer statistical analysis system. **Ciência e Agrotecnologia**, v. 35, n. 6, p. 1039 - 1042, 2011.

FREIRE, J. L. O.; DIAS, T. J.; CAVALCANTE, L. F.; FERNANDES, P. D.; LIMA NETO, A. J. Quantum yield and gas exchange in Yellow Passionflower under water salinity, biofertilizer and mulch. **Revista Ciência Agronômica** v.45, n.1, p.82-91, 2014.

HATTORI, T.; INANAGA, S.; ARAKI, H.; AN, P.; MORITA, S.; LUXOVÁ, M.; LUX, A. Application of silicon enhanced drought tolerance in *Sorghum bicolor*. **Physiologia Plantarum**, v. 123, n. 4, p. 459-466, 2005.

HUANG, Z.; LONG, X.; WANG, L.; KANG, J.; ZHANG, Z.; ZED, R.; LIU, Z. Growth, photosynthesis and H+ ATPase activity in two Jerusalem artichoke varieties under NaCl-induced stress. **Process Biochem, istry**, v.47, n. 4, p. 591-596, 2012.

HASHEMI A, ABDOLZADEH A, SADEGHIPOUR HR. Beneficial effects of silicon nutrition in alleviating salinity stress in hydroponically grown canola, *Brassica napus* L., plants. **Soil Science and Plant Nutrition**, v.56, n. 2, p.244-253, 2010.

HUANG, Z.; LONGA, X.; WANGAL, L.; Growth, photosynthesis and H+-ATPase activity in two Jerusalem artichoke varieties under NaCl-induced stress. **Process Biochem, istry**, v.47, n.4, p.591-596, 2012.

JESUS, E.G.; FATIMA, R.T.; GUERRERO, A.C.; ARAUJO, J.L.; BRITO, M.E.B. Growth and gas exchanges of arugula plants under silicon fertilization and water restriction. **Brazilian Journal of Agricultural and Environmental Engineering,** v.22, n.2, p.119-124, 2018.

KIM YH.; KHAN AL.; WAQAS M.; LEE IJ. Silicon regulates antioxidant activities of crop plants under abiotic-induced oxidative stress: a review. **Frontiers in Plant Science**, n. 8, p. 1-7, 2017.

LEITÃO, T. E. M. F. S.; TAVARES, J. C.; RODRIGUES, G. S. O.; GUIMARÃES, A. A.; DEMARTELAERE, A. C. F. Avaliação de mudas de mamão submetidas a diferentes níveis de adubação nitrogenada. **Revista Caatinga**, v. 22, n. 3, p. 160-165, 2009.

LEE, S.K.; SOHN, E.Y.; HAMAYUN, M.; YOON, J. Y.; LEE, I. J. Effect of silicon on growth and salinity stress of soybean plant grown under hydroponic system**. Agroforest Syst**, v.80, p.333-340, 2010.

LIMA, G. S. de; DIAS, A. S.; SOUZA, L. de P.; SÁ, F. V. da S.; GHEYI, H. R.; SOARES, L. A. dos A. Effects of saline water and potassium fertilization on photosynthetic pigments, growth and production of West Indian Cherry. **Environment & Water Journal, v.**13, n.3, p.1-12, 2018.

LUCENA, C. C.; SIQUEIRA, D. L.; MARTINEZ, H. E. P.; CECON, P. R.; Salt stress change chlorophyll fluorescence in mango. **Revista Brasileira de Fruticultura**, v. 34, n. 4, p.1245-1255, 2012.

MAHDIEH, M.; HABIBOLLAHI, N.; AMIRJANI, M.R.; ABNOSI, M.H.; GHORBANPOUR, M. Exogenous silicon nutrition ameliorates salt-induced stress by improving growth and efficiency of PSII in *Oryza sativa* L. cultivars. **Journal of soil science and plant nutrition**, v.15, n.4, 2015.

MEDEIROS, S. S.; CAVALCANTI, A. M. B.; MARIN, A. M. P.; TINÔCO, L. B. M.; SALCEDO, I. H.; PINTO, T.F. **Sinopse do censo demográfco para o semiárido brasileiro.** Campina Grande, INSA, 2012, 103 p.

MELETTI, L. M. M. Avanços na cultura do maracujá no Brasil. **Revista Brasileira de Fruticultura**, v. 33, n. 1, p. 83-91, 2001.

MUNNS, R.; TESTER M. Mechanisms of salinity tolerance. **Annual Review of Plant Biology**, v. 59, p. 651-681, 2008.

REIS, M. V.; FIGUEIREDO, J. R. M.; PAIVA, R.; SILVA, D. P. C.; FARIA, C. V. N.; ROUHANA, L. V. Salinity in rose production. **Ornamental Horticulture**, v.22, n.2, p. 228-234, 2016.

REZENDE, R. A. L. S.; RODRIGUES, F. A.; SOARES, J. D. R.; SILVEIRA, H. R. O.; PASQUAL, M.; DIAS, G. M. G. Salt stress and exougnous silicion influence physiological and anatomical features of in vitro-grown cape gooseberry. **Rural Science**, v.48, n.1, p.1-9, 2017.

RICHARDS, L. A. **Diagnosis and improvement of saline and alkali soils**. Agriculture Handbook No. 60, Washington: USDA, Department of Agriculture, 1954. 160 p.

SÁ, F. V. S. **Ecophysiology of acerola irrigated with saline water under nitrogen and phosphorus doses**, UFCG, 2018. 150p. Thesis (PhD) - Federal University of Campina Grande, Campina Grande, PB.

SAIRAM, R.K.; TYAGY, A. Physiology and molecular biology of salinity stress tolerance in plants. **Current Science**, v. 86, n. 3, p. 407-421, 2004.

SANTOS, J. B. **Estudo das relações nitrogênio: potássio e calcio: magnésio sobre o desenvolvimento vegetativo e produtivo do maracujazeiro amarelo** / João Batista Santos. - Areia, 2001. 88f. : il, color.

SILVA, R. M. **Produção de mudas de maracujazeiro-amarelo com diferentes tipos de enxertia e uso de câmera úmida,** UFERSA, 2012. 59p. Dissertation (Master) Universidade Federal Rural do Semi-árido, Mossoró, RN.

SIVANESAN, I.; SON, M. S.; LIM, C. S.; JEONG, B. R. Effect of soaking of seeds in potassium silicate and uniconazole on germination and seedling growth of tomato cultivars, Seogeon and Seokwang. **African Journal of Biotechnology**, v. 10, n. 35, p. 6743-6749, 2011.

TAIZ, L.; ZEIGER, E; MOLLER, I. M.; MURPHY, A. **Fisiologia e desenvolvimento vegetal**. 6.ed. Porto Alegre: Artmed, 2017. 918p.

TAIZ, L.; ZEIGER, E. **Fisiologia vegetal**. 5. ed. Porto Alegre, Artmed, 2013. 918 p.

TAHIR, M.A.; AZIZ, T.; FAROOQ, M., SARWAR, G. Silicon-induced changes in growth, ionic composition, water relations, chlorophyll contents and membrane permeability in two salt-stressed wheat genotypes. **Archives of Agronomy and Soil Science**, v.58, n.3, p.247-56, 2012.

WANDERLEY, J.A.C.; AZEVEDO, C.A.V.; BRITO, M.E.B.; CORDÃO, M.A.; LIMA, R.F.; FERREIRA, F.N. Nitrogen fertilization to attenuate the damages caused by salinity on yellow passion fruit seedlings. **Brazilian Journal of Agricultural and Environmental Engineering**, v.22, n.8, p.541-546, 2018.

CHAPTER II

QUALITY OF YELLOW PASSION-FRUIT SEEDLINGS UNDER SALT STRESS AND SILICATE FERTILISATION

Resumo: Na região semiárida do Nordeste Brasileiro ocorre longos períodos de estiagem, destacando-se como fator limitante para produção agrícola e com a escassez qualitativa e quantitativa, o uso de águas salinas surge como alternativa para expansão das áreas irrigadas, associado a utilização destas águas a adubação silicatada surge como um importante atenuador do estresse salino provocado pela utilização de águas com elevadas CEa. Neste sentido, objetivou-se com este trabalho avaliar a produção de fitomassas e o índice de qualidade de Dickson de maracujazeiro amarelo cultivadas sob salinidade crescente da água e adubação silicatada. The experiment was developed in vegetation house conditions in the Center of Science and Technology Agroalimentary in the city of Pombal-PB. The experimental design was based on randomized block design in a 5 x 5 factorial scheme, with five levels of electrical conductivity of the irrigation water (0.3; 1.0; 1.7; 2.4 and 3.1 dS m-1) and five doses of silicate fertilization (0; 25; 50; 75 and 100 g of potassium silicate plant-1), with four repetitions and two plants per plot, totalizing two hundred experimental units. The accumulation of phytomass (leaves, stem and root), total dry phytomass, aboveground dry phytomass, root/aboveground ratio and Dickson's quality index were evaluated. The data were subjected to the F test at 0.01% and 0.05% probability, when there was a significant effect of treatments, the means were subjected to polynomial regression for salt levels and doses of silicon. The conductivity of the water from 0.3 dS m-1 caused a decrease in the production of phytomass of seedlings of yellow giant passion-fruit. However, according to the results of the Dickson quality index it is possible to produce quality seedlings of passion-fruit with water salinity up to 3.1 dS m-1. The doses of silicate fertilization softened the effect of salt stress on the root/head ratio of the plants of yellow giant passion-fruit.

Keywords: saline waters, attenuating agent, silicon.

Abstract: In the semiarid region of Brazilian Northeast, long periods of drought occur, standing out as a limiting factor for agricultural production and with the qualitative and quantitative scarcity, the use of saline water appears as an alternative for the expansion of irrigated areas. Associated to the use of those waters, silicate fertilization appears as an important attenuator of salt stress caused by the use of waters with high CEa. In this sense, this work aims to evaluate the production of phytomass and Dickson's quality index of yellow passion fruit cultivated under increasing salinity of the water and silicate fertilization. The experiment was carried out under greenhouse conditions at the Center for Science and Agri-Food Technology in the municipality of Pombal-PB. The experimental design was used in randomized blocks 5 x 5 factorial scheme, referring to five levels of electrical conductivity of the irrigation water (0.3; 1.0; 1.7; 2.4 and 3.1 dS m-1) and five doses of silicate fertilization (0; 25; 50; 75 and 100 g of potassium silicate plant-1) with four replications and two plants per plot, totalizing two hundred experimental units. The accumulation of phytomass (leaves, stem and root), total dry phytomass, dry shoot phytomass, root / shoot ratio and Dickson's quality index were evaluated. The obtained data were submitted to the F test at 0.01% and 0.05% probability. When there was a significant effect of the treatments, the averages were submitted to polynomial regression for the saline levels and for the doses of silicon. The conductivity of the water from 0.3 dS m-1 caused a decrease in the production of phytomass of passion fruit seedlings. However, according to the result of the Dickson quality index, it is possible to produce seedlings of passion fruit quality with salinity of water up to 3.1 dS m-1. The doses of silicate fertilizer mitigated the effect of saline stress on the root / shoot ratio of the giant yellow passion fruit plants.

Key words: saline water, attenuating, silicon.

1 INTRODUCTION

The passion-fruit originated in Tropical America, containing more than 150 species native to Brazil (GONÇALVES; SOUZA, 2006). The yellow passion-fruit (*Passiflora edulis* f. flavicarpa) is grown in 95% of the commercial area and has a high economic and social value, being associated with human nutrition (CANÇADO JUNIOR et al., 2000; MELETTI, 2003).

Species of Passiflora are cultivated mainly for the nutritional characteristics of its fruit and pharmacological properties (sedative, diuretic, antidiarrheal, stimulant, tonic and in the treatment of hypertension) and food of its juice, seed and peel, being rich in pectin, niacin, iron, calcium and phosphorus. However the main importance of passion fruit is in human nutrition, and it can be consumed *in natura* or in the processing of juices, soft drinks, jams, jellies, ice cream and liqueurs (COELHO et al., 2016).

Although the north-east region of Brazil has the largest production of yellow passion-fruit, the excess of salts in the springs has severely compromised the formation of seedlings of the crop (AYERS; WESTCOT, 1999). The effects of irrigation water salinity on the plants are reflected in alterations of the osmotic potential, ion toxicity and nutritional unbalance of the plants (AZEVEDO NETO; TABOSA, 2000; FERREIRA et al., 2007). According to Ayers & Westcot (1999) passion-fruit is a crop sensitive to saline stress with a threshold salinity of 1.3 dS m-1. Salinity hinders plant growth because of osmotic effects, restricting the availability and absorption of nutrients; by toxicity with the accumulation of specific ions, mainly Na+ and Cl- and by disturbances in plant nutrition, directly affecting the metabolism and growth of plants (FREIRE et al., 2010).

There are several works (OLIVEIRA et al., 2015; CAVALCANTI et al., 2009; MOURA et al., 2017) developed with the culture of passion-fruit under salinity conditions, however there are incipient researches evaluating the formation of seedlings under salt stress conditions and silicate fertilization. Considering the importance of silicon for the maintenance of the photosynthetic rate, increase of the stomatal conductance of the plant, decrease of the transpiration rate through the control of the mechanism of opening and closing of stomas, improvement of the architecture of the plants, among others. Besides being associated with the increase of the capacity of antioxidant defense in several species of plants, increasing the activity of enzymes associated with the defense of plants against water deficiency and salt stress (GONG et al., 2005; HATTORI et al., 2005). The use of (Si) can be used as an alternative to minimize the negative effect of salts on crops (LIMA et al., 2011).

In this context, the objective was to evaluate the phytomass and Dickson's quality index in the formation of yellow passion-fruit seedlings as a function of irrigation with saline water and silicate fertilization.

2 MATERIAL AND METHODS

2.1 Site of the experiment

The experiment was conducted under greenhouse conditions in the Centre for Agro-Food Sciences and Technology of the Federal University of Campina Grande, in the municipality of Pombal-PB, located at 6°47'3" S, 37°49'15" W and 193 m altitude.

2.2 Experimental Procedure

A randomized block design was adopted, with treatments arranged in a 5 x 5 factorial scheme relating to five levels of electrical conductivity of the irrigation water - CEa (0.3; 1.0; 1.7; 2.4 and 3.1 dS m-1) and five doses of silicate fertilization (0; 25; 50; 75 and 100 g of potassium silicate per plant) with two plants per plot and four repetitions, totaling two hundred experimental units. The salinity levels of the water were determined based on research developed by Andrade (2018). The levels of ECa were obtained from the addition of sodium chloride (NaCl) in the water supply of the city of Pombal-PB (0.3 dS m-1) obeying the relationship between ECa and the concentration of salts (mg L-1 = 640 x ECa) (RICHARDS, 2000).

Potassium silicate was used as a source of silicon. It is a compound of multi-minerals: selenium, vanadium, calcium, zinc, phosphorus and various trace elements, with 50% SiO_2 and 4% K_2O.

The culture which was studied was the Yellow Giant passion fruit, a hybrid adapted to altitudes varying between 376 and 1100m, which can be planted at any time of the year. Intended for the fresh market and industry. Bright yellow fruits, weighing between 120 and 350 grams. The pulp represents 40% of the fruit, with a strong yellow colour and high percentage of vitamin C. °Brix between 13 and 15 degrees.

The research was conducted in plastic bags with dimensions of 15 x 20 cm, with a capacity of 1250 ml, filled with substrate 2:1:1 in volume base (soil, sand and tanned bovine manure). The soil moisture content was raised to field capacity with water of the lowest saline level (0.3 dS m-1), where water was placed in the bags until free drainage occurred. Two seeds were sown per bag and after seedling emergence, only one plant per container was thinned when they were 10 cm high.

Irrigation was performed daily manually using water from the respective treatment and based on the process of drainage lysimetry (Bernardo et al., 2006). The volume applied in each irrigation was determined by the difference between the volume applied and the drained volume of the previous day, plus a leaching fraction of 15% applied every 20 days. The saline levels were applied daily 30 DAS being applied until 60 DAS.

The silicon doses were applied diluted via irrigation water and started at 30 DAS and were applied weekly, totalizing 4 applications until 60 DAS, according to the treatments, where 0, 6.25, 12.5, 18.75 and 25g were applied weekly per plant.

The physical and chemical characteristics of the soil used for the production of seedlings were analysed by the laboratory of irrigation and salinity of UFCG, as shown in the results of (Table 1).

Table 2. Physical and chemical characteristics of the soil used for the production of seedlings of Yellow Passionflower, performed by the laboratory of irrigation and salinity of the UFCG campus of Campina Grande-PB. Pombal-PB, 2019.

				..... Chemical attributes					
Ph		P		K+	Na+	Ca2+	Mg2+	Al3+	H + Al3+
CaCl2 1:2.5	ECes (dS m-1)	mg dm-3		.....cmolc dm-3.....					
7,00	0,20	0,21		0,38	0,09	2,50	3,75	0	0
....		..... Physical Attributes							
Sand		Silt	Clay	Ds	Dp	Porosity	UD	Textural class	
.....g kg-1.....				..... kg dm-3.....		%	%		
85,30		13,07	1,63	1,50	2,69	47,23	0,55	Areia Franca	

pHes = pH of the saturation extract of the substrate; ECes = Electrical conductivity of the saturation extract of the substrate at 25 °C. Ca2+ and Mg2+ extracted with 1 M KCl pH 7.0; Na+and K+ extracted using 1 M NH4OAc pH 7.0; H+ and Al3+ extracted with 0.5 M CaoAc pH 7.0; FA - Clayy loam; AD - Available water. Ds - Soil density, Dp - Particle density. UD - Moisture (dry soil base).

Fertilization with nitrogen and potassium was carried out weekly; for the potassium fertilization the content of the element already present in the potassium silicate was taken into consideration. The fertilization with phosphorus was applied in the foundation according to Santos (2001). It was used as a source of nitrogen the urea, as a source of phosphorus the MAP and as a source of potassium the potassium chloride.

Fertilization description: NPK fertilization was carried out using 100 mg N kg-1 soil, 150 mg K2O kg-1 soil and 300 mg P2O5 kg-1 soil. The following fertilizers were applied per plant: 0.217 g of urea, 1.4 g of monoammonium phosphate and 0.56 g of potassium chloride, divided into 4 portions, supplied via fertigation at 20 days after planting.

The micronutrients were applied weekly at 30 DAS using the product Quimifol (water, nitrogen solution, boric acid, lignosulfonates, amino acids, polysaccharides, leonardite and preservative) applying 0.5 g per litre (Cavalcanti, 2008). During the experiment, the cultivation and phytosanitary treatments recommended for the crop, such as pest control and removal of tendrils, were carried out.

2.3 Variables Analysed

The variables were analyzed at 60 days after sowing (DAS) period in which the seedlings were ready to go to the field, being evaluated through the accumulation of dry phytomass (leaves, stem and root). The plants were cut flush with the ground, separated into leaves, stem and roots and placed separately in paper bags, duly identified and placed to dry in an oven with forced air circulation, maintained at a temperature of 65°C until constant mass was obtained, when they were then weighed on precision scales, determining leaf dry phytomass (FSF), stem dry phytomass (FSC) and root dry phytomass (FSR), the sum of which resulted in total dry phytomass (FST).

The aboveground dry phytomass (FSPA), root/aboveground ratio (R/PA) and the Dickson quality index (IQD) were also determined. The FSPA was obtained by adding FSF + FSC. The Dickson quality index (DQI) for seedlings was determined using the formula of Dickson et al. (1960), described by the equation below:

$$IQD = \frac{(FST)}{(AP/DC) + (FSPA/FSR)}$$

where:

IQD = Dickson quality index, AP = plant height (cm), CD = stem diameter (mm), FST = total plant dry phytomass (g planta-1), FSPA = aboveground plant dry phytomass (g planta-1) and FSR = root dry phytomass (g planta-1).

2.4 Statistical analysis

The data obtained in this research were submitted to the F test at 0.01 and 0.05% probability, to carry out variance analysis. When there was a significant effect of the treatments, the means of the variables were submitted to polynomial regression for the doses of silicon and salt levels. The statistical analyses were performed using the software SISVAR Version 5.6 (FERREIRA, 2011).

3 RESULTS AND DISCUSSION

According to the analysis of variance, there was a significant effect of the salt levels (Table 2) on the dry leaf phytomass (FSF), stem (FSC), root (FSR) and aboveground phytomass (FSPA) of passion-fruit plants at 60 DAS. The silicon doses affected significantly only the dry phytomass of the stem (FSC). There was no interaction between the NS x DS factors for any of the variables analyzed

Table 2. Summary of the analysis of variance for leaf (FSF), stem (FSC), root (FSR) and aboveground (FSPA) dry phytomass of passion-fruit 'Gigante amarelo' cultivated under different levels of salinity of irrigation water and doses of silicon, 60 days after sowing. Pombal, CCTA/UFCG, 2020.

Source of variation	GL	Mean Squares			
		FSF	FSC	FSR	FSPA
Salt levels (NS)	4	5,21*	1,30*	0,76**	11,02*
Linear Reg.	1	19,41*	4,63*	2,70*	42,94*
Quadratic Reg.	1	0.78ns	0.50ns	0,28ns	0.03ns
Silicon dose (DS)	4	0.43ns	0,63**	0.17ns	0.72ns
Linear Reg.	1	2,59**	0,87**	1,59*	6,46*
Quadratic Reg.	1	0,41ns	0,19ns	0,21ns	0,04ns
Interaction (NS x DS)	16	1,28ns	0,42ns	0,26ns	2.55ns
Blocks	3	0.31ns	0,05ns	0.09ns	0.13ns
Waste	72	0,39	0,16	0,13	0,59
CV (%)		23,18	21,17	22,33	16,64

ns, **, *, respectively not significant, significant at $p < 0.01$ and $p < 0.05$.

The dry phytomass of leaves (Figure 1A) decreased linearly with the increase in water salinity, obtaining a decline of 12.90% per unitary increment of ECa. Comparing the dry phytomass of leaves of the plants irrigated with an ECa of 3.1 dS m-1 with those that were subjected to the lowest salinity level (0.3 dS m-1) there was a reduction in the FSF of 36.13%. The decrease in the production of phytomass may be associated with changes in photosynthetic rate, where there is a diversion of energy that is destined for growth to the activation and maintenance of metabolic activity associated with salinity tolerance mechanisms such as the maintenance of membrane integrity and the regulation of ionic transport and distribution in various organs within the cells (SOUZA et al., 2016).

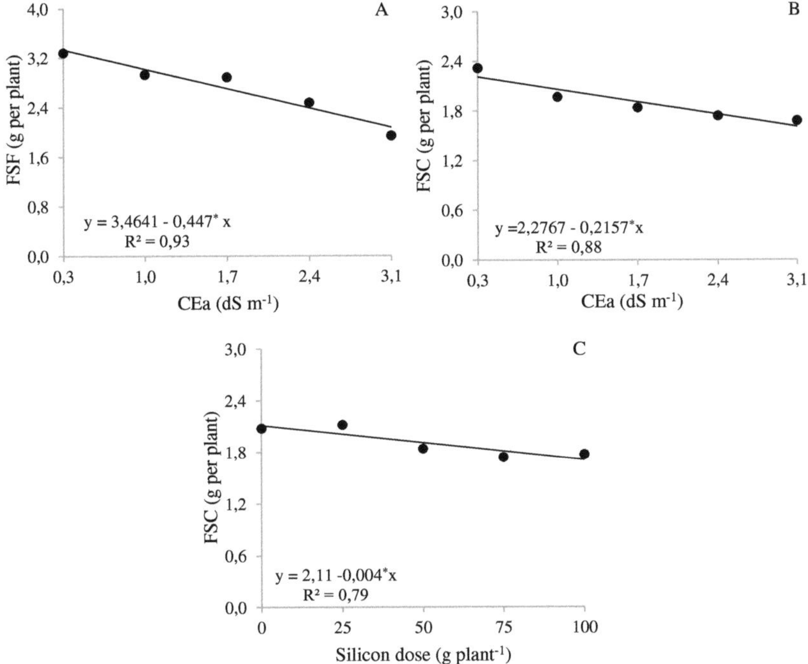

Figure 1. Leaf dry phytomass - FSF (A) stem dry phytomass - FSC (B) in passion-fruit plants 'Gigante Amarelo' as a function of water salinity levels - CEa and stem dry phytomass - FSC (C) as a function of silicon doses, at 60 days after sowing. Pombal, CCTA/UFCG, 2020.

Similarly, the dry phytomass of the stem (Figure 1B) had a linear decreasing behavior as the salinity levels of the irrigation water increased, with a decrease of 9.47% per unit increase in the ECa, that is, a decrease of 0.60 g plant-1 in the plants that received the highest saline level (3.1 dS m-1) in relation to those that were grown under the lowest saline level (0.3 dS m-1). Similar results were found in a study by Nascimento et al. (2017)

with passion-fruit culture subjected to ECa of 0.43 and 4.5 dS m-1, when they observed a reduction in stem dry phytomass from 0.7 g to 0.1 g, indicating a reduction of 86% when increasing the electrical conductivity of irrigation water from the lowest to the highest saline level. Cavalcanti et al. (2009b), in a study of yellow passion-fruit irrigated with saline water, found that the dry phytomass of the stem was reduced with increasing salinity levels of irrigation water.

The silicon doses negatively influenced the dry phytomass of the stem (Figure 1C), with a reduction of 4.73% for each increase of 25 g, that is, with the increase in the doses of Si from 0 to 100 g plant-1 there was a decrease of 1.71 g plant-1 in the dry phytomass of the stem. Linhares (2019) studying sources of silicon in passion-fruit culture, found that the sources of silicon did not provide increases for the dry mass of the stem for *Passiflora edulis.* This situation may be related to the way in which silicon is accumulated, especially in the leaf, which is why it is found in smaller quantities in the stem of the plant.

The dry phytomass of the roots (Figure 2A) of the yellow passion-fruit plants decreased linearly, decreasing as the salinity of the irrigation water increased, with a reduction of 10.72% per unit increase in the ECa. When compared in relative terms, the plants grown under water salinity of 3.1 dS m-1 had a decline in FSR of 0.47 g plant-1 in relation to those grown under the lowest salinity level (0.3 dS m-1). The decrease in the dry phytomass of the roots may be related to the restriction in the elongation of the roots, standing out as a mechanism of tolerance of the genotype itself, reducing the absorption of water and consequently of salts, in order to mitigate the toxicity (MOURA et al., 2017). According to Azevedo et al. (2017) the deposition of soluble salts, such as chloride and sodium when found in excess in the soil or irrigation water causes physiological disturbances in crops.

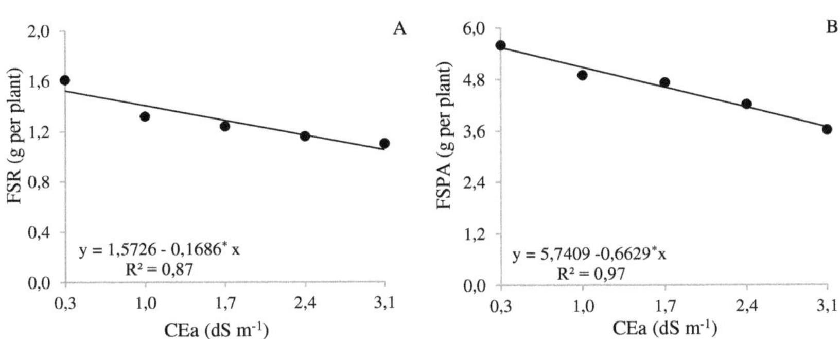

Figure 2. dry phytomass of the root - FSR - (A) and aboveground - FSPA - (B) in passion-fruit plants 'Gigante Amarelo' as a function of water salinity levels - CEa, at 60 days after sowing. Pombal, CCTA/UFCG, 2020.

The dry phytomass of the aerial part of passion-fruit plants was also significantly affected by the salinity of the water, occurring linear decrease of 11.54% per unitary increment of the CEa, with reduction of 32.23% in the FSPA of the plants irrigated with CEa of 3.1 dS m-1 in relation to those that were cultivated under salinity of water of 0.3 dS m-1 . Santos et al. (2016) report that the effect of the increase in salt concentration causes a reduction in the part area of the plants, because they do not have an osmotic adjustment as an adaptation mechanism to the excess salts in the soil solution. Santos et al. (2013) report that the reduction in the production of phytomass in plants may be associated with the ionic and/or osmotic components of salt stress. With the low availability of water there is a mechanism for the closure of the stomata, reducing CO_2 assimilation, a fact that directly affects the production of phytomass of the plants.

Similarly Andrade et al. (2018) in research with the culture of yellow passion fruit subjected to different salinities of irrigation water (0.7; 1.4; 2.1 and 2.8 dS m-1), found decreases by increasing the ECa in the order of 17.86%; 14.10% and 16.10% for the variables leaf dry phytomass, stem dry phytomass and aboveground dry phytomass, at 205 days after transplanting.

According to the variance analysis, there was a significant effect of the saline levels on the root to aboveground ratio (R/AP), the total dry phytomass (FST) and the Dickson quality index (IQD) of passion-fruit plants, at 60 DAS. The silicon doses affected significantly only the root to aerial part ratio (R/AP). There was no interaction between the factors (NS x DS) on any of the variables analyzed, at 60 days after sowing. (Table 3).

Table 3. Resumo da análise de variância para relação raiz/ seca parte aérea (R/PA), fitomassa seca total (FST) e o índice de qualidade de Dickson (IQD) de plantas de maracujazeiro 'Gigante amarelo' cultivado sob diferentes níveis de salinidade da água de irrigação e doses de silício, aos 60 dias após a sem sedura. Pombal, CCTA/UFCG, 2020.

Source of variation	GL	Mean Squares		
		R/PA	FST	IQD
Salt levels (NS)	4	0,16*	17,31*	0,03*
Linear Reg.	1	0,63*	67,33*	0,14*
Quadratic Reg.	1	0.01ns	0.12ns	0.00001ns
Silicon dose (DS)	4	0,07**	0.94ns	0.001ns
Linear Reg.	1	0,34*	14,49*	0,02**
Quadratic Reg.	1	0,04ns	0.06ns	0.003ns
Interaction (NS x DS)	16	0.03ns	3.50ns	0.006ns
Blocks	3	0,41ns	0.09ns	0.007ns
Waste	72	0,01	0,64	0,005
CV (%)		24,49	13,58	20,29

ns, **, *, respectively not significant, significant at $p < 0.01$ and $p < 0.05$.

The root/head ratio of passion-fruit plants (Figure 3A) was significantly affected by the salinity of the water, with a linear increase as the salinity levels of the irrigation water increased, which increased by 36.08% per unit increase in the ECa, that is, an increase of 101.02% (0.52 g plant-1) in plants that received the highest salinity level (3.1 dS m-1) relative to those that received the lowest salinity level (0.3 dS m-1).

This result shows that the aerial part of the plants is more sensitive to increased salinity than the root system. This is important for the optimization of the process of water and nutrient absorption, since salinity causes stresses that limit plant growth and development. Since the reduction of growth parameters is the result of defense strategies, such as the reduction of leaf and stem area and aerial abscission (TAIZ; ZEIGER 2017).

In a study by Cavalcante et al. (2009) with the culture of yellow passion-fruit irrigated with saline water with concentrations from 0.4 to 4.0 dS m-1, it was found that the R/PA ratio decreased with the increase in the electrical conductivity of the irrigation water.

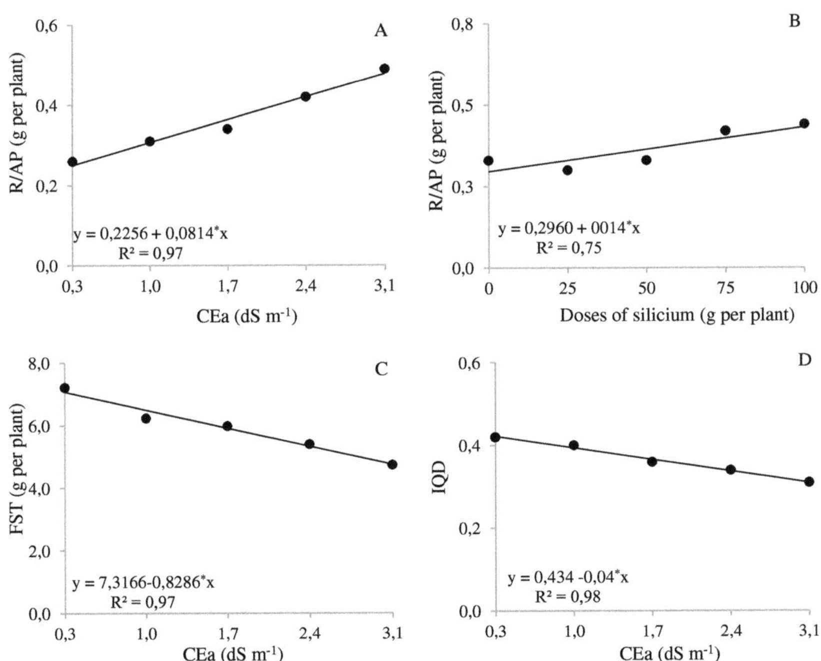

Figure 3. root/head ratio (R/AP) - as a function of water salinity levels - CEa (A) and silicon doses (B), total dry phytomass (FST) (C) and the Dickson quality index (IQD) - (D) in passion-fruit plants 'Gigante Amarelo' as a function of water salinity levels - CEa, at 60 days after sowing. Pombal, CCTA/UFCG, 2020.

Regarding the effect of silicic acid fertilization on the R/AP ratio (Figure 3B), a linear increasing behavior was observed, with an increase of 11.82% for each increase of 25 g plant-1 in the silicon dose. The increase in the root to aerial part ratio presents itself as a strategy of tolerance for the plant, since when they are subjected to stresses they seek to increase their roots in search of water and nutrients (METHRABANJOUBANI et al., 2015) when evaluating the effect of silicon application on cotton, canola and wheat seedlings and found that all plants grown in the presence of Si, at a dose of 1.5 mmol L-1 Si, for 12 days, showed longer roots when compared with plants in the absence of Si.

Fawe et al. (2001) indicate that silicon in the roots plays a role in the signaling network and may induce systemic resistance in other organs. A study conducted by Souza (2015) with passion-fruit trees showed that passion-fruit trees accumulate silicon primarily in the roots, rather than in the leaves and stem. Where MA et al. (2001) reports that values above 1.0 are considered accumulators, between 1.0 and 0.5 are considered intermediate and less than 0.5 are considered non-accumulators.

For the total dry phytomass of passion-fruit (Figure 3C) there was a linear decrease with the increase of the water salinity levels, occurring a reduction of 11.32% per unitary increase of ECa, that is, a decline in the TDF of 31.70% (2.32 g plant-1) in the plants cultivated under the highest salinity level (3.1 dS m-1) in relation to those that received the lowest salinity level (0.3 dS m-1). The decrease in total dry phytomass under salt stress conditions may be associated with the fact of the attempt of osmotic adjustment of the plant, occurring a release of energy for the accumulation of sugars, organic acids and ions in the vacuole, energy that would be used for the accumulation of phytomass in the plant (SANTOS et al., 2012). Corroborating the results obtained in this study, Andrade et al. (2018) in a study evaluating the growth of yellow passion-fruit according to the salinity of the irrigation water CEa (0.7; 1.4; 2.1 and 2.8 dS m-1), verified a linear decrease in the accumulation of dry phytomass of leaves, stem and total of 113.64; 78.78 and 192.41g plant-1 when subjected to the highest saline level.

For the Dickson index of quality (Figure 3D), a linear decreasing behavior was observed as a function of the increase in the salinity of the water, with a reduction of 9.21% per unitary increase of the CEa, that is, the passion-fruit plants irrigated with CEa of 3.1 dSm-1 had a reduction in IQD of 25.80% plants in comparison with those that received the lowest level (0.3 dS m-1). Despite the reduction of the IQD with the increase of the salinity of the water, the seedlings of passion-fruit when submitted to the CEa of 3.1 dS m-1 presented an IQD of 0.31. This fact is important from the point of view that, even under conditions of saline stress the seedlings of passion-fruit had an IQD superior to 0.2 being considered as of good final quality for establishment in the field according to the criteria established by GOMES et al. (2003). Corroborating the results found in the present work, Moura et al. (2017) in a study carried out with seedling formation of species of the *Passiflora* genus under salt stress (0.3; 1.4; 2.5; 3.6 and 4.7 dS m-1), obtained IQD with a value above 0.2 when the plants were irrigated with an EC of 2.5 dS m-1.

4 CONCLUSIONS

Salt stress starting at 0.3 dS m-1 causes a reduction in the production of phytomass of passion-fruit at 60 days after sowing, with the dry phytomass of the leaves standing out as the most sensitive variable. Despite the reduction in the accumulation of phytomass, it is possible to produce passion-fruit seedlings with water salinity of 3.1 dSm-1 with a Dickson quality index of 0.31 being considered acceptable.

The doses of silicate fertilization resulted in the highest root to aerial part ratio of the plants of yellow giant passion-fruit.

5 BIBLIOGRAPHICAL REFERENCES

ALVES, F. A. L.; FERREIRA-SILVA, S. L.; SILVEIRA, J. A. G.; PEREIRA, V. L. A. Efeito do Ca2+ externo no conteúdo de Na+ e K+ em cajueiros expostos a salinity. **Revista Brasileira de Ciências Agrárias**, v.6, n.4, p. 602-608, 2011.

AYERS, R.S.; WESTCOT, D.W. **A qualidade da água na agricultura**. Campina Grande: UFPB, 1999. (FAO Studies: Irrigation and Drainage, 29, Review).

ANDRADE, J.R.; SOUSA, A.; JUNIOR, M.; OLIVEIRA, S.; PAULA; REZENDE, L. P.; ARAÚJO NETO, J. C. Germination and morphophysiology of passion fruit seedlings under salt water irrigation. **Tropical Agricultural Research**, v. 48, n. 3, p. 229-236, 2018.

ANDRADE, E. M. G. **Saline waters and foliar application of hydrogen peroxide in the cultivation of yellow passion fruit 2018**. UFCG, 2018, 104p. Thesis (Doctoral) - Center for Technology and Natural Resources, Federal University of Campina Grande. Campina Grande, PB.

AZEVEDO NETO A. D.; TABOSA J. N. Stress salino em plântulas de milho: Parte II - Distribuição dos macronutrientes catiônicos e suas relações com o sdio. **Revista Brasileira de Engenharia Agrícola e Ambiental**, v.4, n.2, p.165-171, 2000.

AZEVEDO, P. R. L; BEZERRA, D. E. L; SOUTO, F. M; BITU, S. G; PEREIRA JUNIOR, E. D. Efeito dos salais e da qualidade da água no solo e na planta. **Revista de Agroecologia no Semiárido**, v. 1, n. 1, p. 01-12, 2017.

ANDRADE, E. M. G.; LIMA, G. S.; LIMA, V. L. A.; SILVA, S. S.; GHEYI, H. R.; SILVA, A. A. R. Gas exchanges and growth of passion fruit under saline water irrigation and H2O2 application. **Revista Brasileira de Engenharia Agrícola e Ambiental**, v. 23, n. 12, p. 945-951, 2019.

BERNARDO, S; SOARES, A. A.; MANTOVANI, E. C. **Manual de irrigação**. Viçosa: UFV, 2006. 625p.

CAVALCANTE, L. F.; SOUSA, G. G.; GONDIM, S. C.; FIGUEIREDO, F. L.; CAVALCANTE, Í. H. L.; DINIZ, A. A. Initial growth of yellow passion-fruit tree managed in two substrates irrigated with saline water. **Irriga**, v. 14, n. 4, o.504- 517, 2009b.

CAVALCANTI, J. C. P. **Recomendações de adubação para o estado Pernambuco (2ªaproximação)**. 3.ed. Recife: Instituto agronômico do Pernambuco - IPA, 2008. 212p.

CAVALCANTE, L.F.; SANTOS, G.D.; OLIVEIRA, F.A.; CAVALCANTE, I.H.L.; GONDIM, S.C.; CAVALCANTE, M.Z.B. Crescimento e produção do maracujazeiro amarelo em solo de baixa fertilidade tratado com biofertilizantes líquidos. **Revista Brasileira de Ciências Agrárias**, v.2, n.1, p.15-19, 2007b.

CAVALCANTE, L. F.; SOUSA, G. G.; GONDIM, S. C.; FIGUEIREDO, F. L.; CAVALCANTE, I. H. L.; DINIZ, A. A. Initial growth of yellow passionfruit managed on two substrates irrigated with saline water. **Irriga,** v. 14, n.4, p. 504-517, 2009.

CAVICHIOLI, J. C.; MELETTI, L. M. M.; NARITA, N. Aspectos da Cultura do Maracujazeiro no Brasil. TodaFruta, Jaboticabal, 11 p. 2018. Available at:. Accessed 20 Dec. 2019.

Coelho, E.M.; Azêvedo, L.C.; Umsza-Guez, M. (2016) Passion fruit: economic and industrial importance, production, by-products and technological prospection. *Cad. Prospec.* - Salvador - BA, 9 (3): 347-361.

CANÇADO JUNIOR, F. L. et al. de. Aspectos econômicos da cultura do maracujá. Informe **Agropecuário**, Belo Horizonte, v.21, n. 206, p. 10-17, 2000.

DICKSON, A.; LEAF, A. L.; HOSNER, J. F. Quality appraisal of white spruce and white pine seedling stock in nurseries. **The Forest Chronicle**, v. 36, n. 01, p. 10-13, 1960.

FAWE, A.; MENZIES, J.G.; CHERIF, M.; BELANGER, R.R. Silicon and disease resistance in dicotyledons. In: DATNOFF, L.E.; SNYDER, G.H.; KORNDÖRFER, G.H. editors. **Silicon in agriculture.** The Netherlands: Elsevier Science; p. 159-169, 2001.

FERREIRA, P. A.; et al. Relative corn yield and foliar nitrogen, phosphorus, sulfur and chloride contents as a function of soil salinity. **Revista Ciência Agronômica**, v.38, n.1, p.7-16, 2007.

FERREIRA, D. F. SISVAR: A computer statistical analysis system. **Ciência e Agrotecnologia**, v. 35, n. 6, p. 1039 - 1042, 2011.

FIGUEREDO, L. F.; JÚNIOR, S. O. M.; FERRAZ, R. L. S.; DUTRA, A. F.; BEZERRA, J. D.; MELO, A. S. Growth and seedling dry mass partitioning in papaya under salt stress. **Brazilian Journal of Irrigated Agriculture**, v. 12, n. 6, p 2984- 2990, 2018.

FREIRE, A. L. O.; SARAIVA, V. P.; MIRANDA, J. R. P.; BRUNO, G. B. Crescimento, acúmulo de íons e produção de tomateiro irrigado com água salina. **Semina: Agrarian Sciences**, v. 31, supplement 1, p. 1133-1144, 2010.

GONÇALVES, J. S.; SOUZA, S.A.M. Fruta da paixão: panorama of passion fruit economics in Brazil. **Economic Information**, v.36, n.12, 2006.

GONG, H.; ZHU, X.; CHEN, K.; WANG, S.; ZHANG, C. Silicon alleviates oxidative damage of wheat plants in pots under drought. **Plant Science**, v. 169, n. 2, p. 313-321, 2005.

GOMES, J. E. M.; COUTO, L.; LEITE, H. G.; XAVIER, A.; GARCIA, S. L. R. Crescimento de mudas de *Eucalyptus grandis* em diferentes tamanhos de tubetes e fertilização N-P-K. **Revista Árvore**, v. 27, n.2, p. 113-127, 2003.

HATTORI, T.; INANAGAA, S.; ARAKI, H.; NA, P.; MORITA, S.; LUXOVA, M.; LUXE, A. Application of silicon enhanced drought tolerance in Sorghum bicolor. **Physiologia Plantarum**, v. 123, p. 459-466, 2005.

LINHARES, G. A. N. **Silicon sources in papaya and passion fruit: growth, physiology and resistance induction 2019**. Thesis (PhD in plant production) - Centro de Ciências e Tecnologias Agropecuárias, Universidade Estadual do Norte Fluminense Darcy Ribeiro. p. 107. 2019.

LIMA, M. A.; CASTRO, V. F.; VIDAL, J. B. ENEASFILHO, J. Aplicação de silício em milho e feijão-de-corda sob estresse salino. **Revista Ciência Agronômica**, v. 42, n. 2, p. 398-403, 2011.

MA, J. F.; MIYAKE, Y.; TAKAHASHI, E. Silicon as a beneficial element for crop plant. In: DATNOFF, L. E.; KORNDÖRFER, G. H.; SNYDER, G. (Ed.). **Silicon in agriculture,** New York: Elsevier Science, 2001. p. 17-39.

METHRABANJOUBANi, P.; ABDOLZADEH, A.; SADEGHIPOUR, H.R.; AGHDASI, M. (2015) Silicon Affects Transcellular and Apoplastic Uptake of Some Nutrients in Plants. Pedosphere, 25 (2):192-201.

MELETTI, L. M. M. Comportamento de híbridos e seleção de maracujazeiro (passifloraceae). In: SIMPÓSIO BRASILEIRO SOBRE A CULTURA DO MARACUJAZEIRO, 6., 2003. Campos dos Goytacazes. **Proceedings**. Campos dos Goytacazes: Cluster Informática, 2003.

MOURA, R. S.; GHEYI, H. R.; FILHO, C. A.; JESUS, O. N.; LIMA, L. K.; CRUZ, C. S. Formation of seedlings of species from the genus *Passiflora* under saline stress. **Bioscience Journal.** V. 33, n.5, p. 1197-1207, 2017.

NASCIMENTO, E. S.; CAVALCANTI, L. F.; GONDIM, S. C.; SOUZA, J. T. A.; BEZERRA, F. T.; BEZERRA, M. A. F. Formação de mudas de maracujazeiro amarelo irrigadas com águas salinas e biofertilizantes de esterco bovino. **Revista Agropecuária Técnica.** V. 38, n. 1, p. 1-8, 2017.

OLIVEIRA, F. A., LOPES, M. A. C., SÁ, F. V. S., NOBRE, R. G., MOREIRA, R. C. L. PAIVA, E. P. Interaction between salinity of irrigation water and substrates in the production of yellow passion-fruit seedlings. **Comunicata Scientiae**, v. 6, n. 4, p. 471-478, 2015.

RICHARDS, L. A. **Diagnosis and improvement of saline and alkali soils**. Agriculture Handbook No. 60, Washington: USDA, Department of Agriculture, 1954. 160 p.

SANTOS, D.; P.; SANTOS, C. S.; SILVA, P. F.; PINHEIRO, M. P. M. A.; SANTOS, J. C. Growth and phytomass of sugar beet under supplemental irrigation with water of different saline concentrations. **Revista Ceres**, v. 63, n. 4, p. 509-516, 2016.

SANTOS, L. A. A.; LIMA, G. S.; NOBRE, R. G.; GHEYI, H. R.; PEREIRA, F. H. F. Fisiologia e acúmulo de fitomassa pela mamoneira submetida a estresse salino e adubação nitrogenada. **Revista Verde de Agroecologia e Desenvolvimento Sustentável**, v.8, n.1, p.247-256, 2013.

SILVA, E. M., NOBRE, R. G., SOUZA, L. P., ARAÚJO, R. H. C. R., PINHEIRO, F. W. A., ALMEIDA, L. L. S. Morfophysiologia de portaenxerto de goiabeira irrigado com águas salinizadas sob doses de nitrogen. **Comunicata Scientiae**, v. 8, n. 1, p. 32-42, 2017.

SOUZA, de P. S.; NOBRE, R. G.; SILVA, E. M de.;LIMA, G. S de.; PINHEIRO; F. W. A.; ALMEIDA; L. L. de S.; Formation of 'Crioula' guava rootstock under saline water irrigation and nitrogen doses. **Revista Brasileira de Engenharia Agrícola e Ambiental**, Campina Grande. v.20, n.8, p.739-745, 2016.

SOUZA, B. N. **Silício no desenvolvimento morfofisiológico de mudas de maracujazeiro amarelo 2015**. Dissertation (Academic Master) - Universidade Federal de Lavras. p. 79. 2015.

TAIZ, L.; ZEIGER, E.; MOLLER, I.; MURPHY, A. **Physiology and plant development**. 6.ed. Porto Alegre: Artmed, 2017. 888 p.

CHAPTER III

PHYSIOLOGICAL INDICES AND GROWTH OF YELLOW PASSION-FRUIT UNDER SALT STRESS AND SILICATE FERTILIZATION

The abiotic stresses are responsible for the loss of agricultural production in different regions of the world, especially in the semiarid regions, which have long periods of drought and high evapotranspiration, which induces the use of saline water as an alternative for the expansion of irrigated areas. Neste sentido, objetivou-se com este trabalho avaliar os índices fisiológicos e o crescimento do maracujazeiro amarelo em função da salinidade da água de irrigação e a adubação silicatada. The experiment was developed in two stages, the first stage consisted in the formation of the seedlings in environment of vegetation house, after sixty days the seedlings were transplanted for pots with capacity for 100 L, beginning thus the second stage in field conditions in the Center of Science and Technology Agroalimentary in the city of Pombal-PB. The experimental design was based on randomized block design in a 5 x 2 factorial scheme, with five levels of electrical conductivity of the irrigation water (0.3; 1.0; 1.7; 2.4 and 3.1 dS m-1) associated with two doses of silicate fertilization (150 and 300 g plant-1 of potassium silicate) with four repetitions and one plant per plot. Salt stress caused a decrease in stomatal conductance, intercellular CO_2 concentration, transpiration, instantaneous water use efficiency and in the growth of passion-fruit 'Gigante Amarelo' plants. The synthesis of chlorophyll *a* was inhibited by the increase of ECa from 0.3 dS m-1. However, the salinity of the water caused an increase in the contents of carotenoids and protoplasmic content, silicon had a significant effect on conductance, photosynthesis, water use efficiency, chlorophyll b, carotenoids and relative growth rate of the diameter.

Key words: *Passiflora edulis* f. flavicarpa, salinity, silicon.

Abstract: Abiotic stresses are responsible for the loss of agricultural production in different regions of the world, especially in semiarid regions, which are exposed to long periods of drought and high evapotranspiration, which induces the use of saline water as an alternative for the expansion of irrigated areas. To soften the stresses caused by those waters, silicate fertilization becomes an important factor to attenuate salt stress. In this sense, this work aims to evaluate the physiological indices and the growth of the yellow passion fruit in function of the salinity of the irrigation water and the silicate fertilization. The experiment was developed in two stages. The first stage consisted of the formation of seedlings in a greenhouse environment. After sixty days, the seedlings were transplanted to 100 L capacity pots, thus the second stage beginning in field conditions at the Center for Science and Agri-Food Technology, municipality of Pombal-PB. A randomized blocks 5 x 2 factorial scheme design was used, whose treatments consisted of five levels of electrical conductivity of the irrigation water (0.3; 1.0; 1.7; 2.4 and 3.1 dS m-1) associated to two doses of silicate fertilization (150 and 300 g plant-1 of potassium silicate) with four replications and one plant per plot. Saline stress caused a decrease in stomatal conductance, intercellular CO2 concentration, transpiration, instant water use efficiency and in the growth of "Giant Yellow" passion fruit plants. The synthesis of chlorophyll a was inhibited by the increase in CEa from 0.3 dS m-1. However, the salinity of the water caused an increase in the levels of carotenoids and protoplasmic content; silicon showed a significant effect on conductance, photosynthesis, water use efficiency, chlorophyll b, carotenoids and relative diameter growth rate.

Key words: Passiflora edulis f. flavicarpa, salinity, silicon.

1 INTRODUCTION

The passion fruit (*Passiflora edulis* f. flavicarpa) is cultivated in almost all Brazilian territory, standing out as the largest producer and consumer, the juice of this fruit is the third most produced by Brazilian agroindustries (FREIRE et al., 2014).

Although the semi-arid region of the Brazilian Northeast has favorable conditions for the cultivation of passion-fruit, its production depends on the use of irrigation because of the seasonality of rainfall, high rates of evapotranspiration and high temperatures favoring water shortage in most months of the year. Salinity affects the availability of water due to changes in the osmotic potential of the soil solution, reducing it to the point that the plant cannot easily extract the water (GHEYI; DIAS; LACERDA, 2010).

The effects of salts on plants can embarrass physiological and biochemical functions causing disturbances in water relations and changes in the absorption and utilization of nutrients available to plants (AMORIM et al., 2010), which may delay their growth and reduce production. Thus, salinity reduces plant growth due to osmotic and toxic effects and losses in the absorption of essential nutrients, affecting gas exchange as a result of water stress caused by the decrease in the osmotic potential of the soil solution (DIAS et al., 2016).

Among the physiological processes affected we highlight photosynthesis, which can be inhibited by the accumulation of $Na+$ and/or $Cl-$ in chloroplasts, affecting the biochemical and photochemical processes that are involved in photosynthesis (TAIZ & ZEIGER, 2009). When the plant is under salt stress condition occurs reduction in stomatal conductance and transpiration, in order to reduce water loss to avoid dehydration (BERTOLLI; SOUZA; SOUZA, 2015). With the decrease of these parameters photosynthesis is impaired (SILVA et al., 2010).

In order to reduce the action caused by salinity, one of the strategies adopted is the use of silicon as a source of fertilization for the plants. It presents several benefits such as: maintenance of the photosynthetic rate, increase of the stomatic conductance of the plant, besides having an antioxidant capacity in the presence of water deficiency (GONG et al., 2005; HATTORI et al., 2005). Studies have shown that the use of silicon can be used as an alternative to minimize the negative effect of salts on crops (LIMA et al., 2011).

In this context, the objective was to evaluate the physiological indices and the growth of yellow passion-fruit, under irrigation with water of different salinities and silicate fertilization.

2 MATERIAL AND METHODS

2.1 Place where the experiment was conducted

The experiment was conducted under field conditions, on the premises of the Centre for Agro-Food Sciences and Technology of the Federal University of Campina Grande, in the municipality of Pombal-PB, located at 6°47'3" S, 37°49'15" W and 193 m altitude.

2.2 Experimental Procedure

A randomized block design was adopted, with treatments arranged in a 5 x 2 factorial scheme relating to five levels of electrical conductivity of the irrigation water (0.3; 1.0; 1.7; 2.4 and 3.1 dS m-1) associated with two doses of silicate fertilization (150 and 300 g per plant of potassium silicate) with four repetitions. The levels of salinity of the water were determined based on research developed by Andrade (2018). They were obtained by adding sodium chloride (NaCl) to the water supply of the city of Pombal-PB (0.3 dS m-1) according to the relationship between ECa and the concentration of salts (mg L-1 = 640 x ECa) (RICHARDS, 2000).

Potassium silicate was used, composed of multiminerals: Selenium, vanadium, calcium, zinc, phosphorus and various trace elements, with 50% SiO2 and 4% K2O as a source of silicon, where 12.5 g per plant were applied at a dose of 150 g, and 25.0 g per plant at a dose of 300 g. The crop studied was the Yellow Giant passion-fruit plant.

The first stage of the research was initiated with sowing in plastic bags with dimensions of 15 x 20 cm, filled with substrate 2:1:1 in volume base (soil, sand and tanned bovine manure). Two seeds were sown per bag and after seedling emergence, only one plant per container was thinned when they were 10 cm high. At 60 DAS they were transplanted into pots, thus beginning the second stage.

The soil moisture content was raised to field capacity with water of the lowest salinity level (0.3 dS m-1). In order to facilitate the process of acclimatization, after transplanting the plants were irrigated with water with EC of 0.3 dS m-1 until 29 days after transplanting (DAT), the application of saline water began after 30 DAT.

Irrigation was performed daily manually using water from the respective treatment and based on the process of drainage lysimetry (BERNARDO et al., 2006). The volume applied in each irrigation was determined by the difference between the volume applied and the drained volume of the previous day, plus a leaching fraction of 15% applied every 20 days and accompanied by the electrical conductivity data of the drained water.

The application of silicon started at 30 DAT and was carried out at intervals of 10 days, totalizing 4 applications until 60 DAT according to the treatments. The different doses of silicon fertilizer were diluted separately in water of the respective treatment.

Figure 1. Experimental view of yellow passion fruit tree seedlings. UFCG, Pombal-PB, 2019.

The seedlings were transplanted into drainage lysimeters (polyethylene buckets) with a capacity of 100 L; each pot was perforated at the base to allow drainage, and coupled to a 4 mm diameter transparent drain. The end of the drain that remained inside the pot was wrapped with a non-woven geotextile blanket (Bidim OP 30) to prevent clogging of the soil material.

The pots were filled with a layer of 0.5 kg of gravel followed by 100 kg of soil material representative of the semi-arid region of the State of Paraíba (duly crushed and homogenized). Below each drain a plastic bottle was attached for the collection of drained water and estimation of water consumption by the plant. Before starting the experiment, the soil was sampled for determination of the chemical and physical-hydric parameters in the Irrigation and Salinity Laboratory (LIS) of the CTRN/UFCG, according to the methodology proposed by Claessen (1997) (Table 1).

Table 3. Physical and chemical characteristics of the soil used for the production of seedlings of Yellow Passionflower, performed by the laboratory of irrigation and salinity of UFCG - Campina Grande-PB campus. Pombal-PB, 2019.

	.... Chemical attributes								
Ph		P		K+	Na+	Ca2+	Mg2+	Al3+	H + Al3+
CaCl2 1:2.5	ECes (dS m-1)	mg dm-3		............cmolc dm-3					
7,00	0,20	0,21		0,38	0,09	2,50	3,75	0	0
....	.... Physical Attributes								
Sand		Silt	Clay	Ds	Dp	Porosity	UD	Textural class	
	..........g kg-1			..... kg dm-3.....		%	%		
85,30		13,07	1,63	1,50	2,69	47,23	0,55	Areia Franca	

pHes = pH of the saturation extract of the substrate; ECes = Electrical conductivity of the saturation extract of the substrate at 25 °C. Ca2+ and Mg2+ extracted with 1 M KCl pH 7.0; Na+and K+ extracted using 1 M NH4OAc pH 7.0; H+ and Al3+ extracted with 0.5 M CaoAc pH 7.0; FA - Clayy loam; AD - Available water. Ds - Soil density, Dp - Particle density. UD - Moisture (dry soil base).

The plants were spaced at 3 m between rows and 2 m between plants, using the vertical espalier system with plain wire #14. When the plants reached 10 cm above the espalier, the apical bud was pruned, aiming at the emission of secondary branches where only two were selected and conducted one to each side until the length of 1.0 m.

After the secondary branches reached the half (1.0 m) of the spacing between plants, new pruning of their apical buds was carried out, aiming the emission of tertiary branches that formed a curtain. Along the conduction of the experiment the elimination of tendrils and thieving branches was carried out, aiming to favour the development of the culture.

Fertilization with nitrogen and potassium was carried out weekly; for the potassium fertilization the content of the element already present in the potassium silicate was taken into consideration. The fertilization with phosphorus was applied in the foundation according to Santos (2001). Ammonium sulphate was used as a source of nitrogen, MAP as a source of phosphorus and potassium chloride as a source of potassium.

12.5 g of ammonium sulphate, 20.83 g of MAP and 8.33 g of potassium chloride were applied per plant via fertigation.

The micronutrients were applied weekly, using the product Quimifol at 0.5 g per litre (CAVALCANTI, 2008). During the experiment, the cultivation and phytosanitary treatments recommended for the crop were carried out.

2.3 Variables analysed

The growth of the passion-fruit was evaluated at 30 days after the beginning of the treatments, by determining the diameter of the stem (CD) using a pachymeter. From these data the absolute growth rate (ACtdc) and relative growth rate (RGRdc) were obtained, adapting procedures contained in (BENINCASA, 2003), according to Equation 1;2.

$$\mathrm{TCAdc} = \frac{(\mathrm{DC_2} - \mathrm{DC_1})}{(\mathrm{T_2} - \mathrm{T_1})} \quad (1) \qquad \mathrm{TCRdc} = \frac{(\ln \mathrm{DC_2} - \ln \mathrm{DC_1})}{(\mathrm{T_2} - \mathrm{T_1})} \quad (2)$$

Where: TCAdc = absolute growth rate of stem diameter (mm day-1), DC1 = stem diameter (mm) at time t1, DC2 = stem diameter (mm) at time t2, TCRdc = relative growth rate of stem diameter (mm mm-1 day-1), ln = natural logarithm.

On the same date, the rate of CO_2 assimilation, transpiration, stomatal conductance, internal CO_2 concentration, instantaneous water use efficiency, instantaneous carboxylation efficiency, chlorophyll content, quantum fluorescence of photosystem II, transpiration and photosynthesis, related to photosynthetically active radiation, were also evaluated in the different treatments.

At 60 days after transplanting, gas exchange was determined through stomatal conductance (*gs*) (mol m-2 s-1), transpiration (*E*) (mmol H2O m-2 s-1), CO2 assimilation rate (*A*) (μmol m-2 s-1) and intercellular CO2 concentration (*Ci*) (μmol mol-1). From these data the instantaneous water use efficiency (*EiUA*) (*A/E*) [(μmol m-2 s-1) (mmol H2O m-2 s-1)$^{-1}$] and the instantaneous carboxylation efficiency (*EiCi)* (*A/Ci*) [(μmol m-2 s-1) (μmol mol-1)$^{-1}$] under photosynthetic photon flux density of 1.200 μmol m-2 s-1, using an infrared gas analyser - IRGA (Infra Red Gas Analyser, model LCpro - SD, from ADC Bioscientific, UK.

The chlorophyll *a*, *b* and carotenoids contents (mg g-1 of fresh matter - MF) were determined following the laboratory method developed by (ARNON, 1949), according to:

Eq. 3, 4 and 5:

$$\text{Chlorophyll } a = (12.7 \times A663 - 2.69 \times A645) \quad (3)$$

$$\text{Chlorophyll } b = (22.9 \times A645 - 4.68 \times A663) \quad (4)$$

$$\text{Carotenoids} = (1000 \times A470 - 1.82 \text{ Chl } a - 85.02 \text{ Chl } b) / 198 \quad (5)$$

In order to evaluate the capacity of cell membrane disruption under salt stress conditions, electrolyte leakage into the cell membrane was determined. Thus, the protoplasmic content was obtained according to Scotti Campos and Thu Pham Thi (1997), according to Eq. 6:.

$$CP = \frac{Ci \times 100}{Cf} \quad (6)$$

where:

CP = protoplasmatic content (%); Ci = initial electrical conductivity (dS m-1); Cf= final electrical conductivity (dS m-1)

2.4 Statistical analysis

The data concerning the measured variables were submitted to the F test at 0.05% and 0.01% of significance, to perform the variance analysis. When there was a significant effect of the treatments, the means of the variables were submitted to the test of Tukey 5% for the doses of silicon and polynomial regression for the saline levels. The statistical analyses were performed using the software SISVAR Version 5.6 (FERREIRA, 2011).

3 RESULTS AND DISCUSSION

Based on the analysis of variance, it was observed that the interactive effect between the saline levels of the irrigation water and doses of silicon on the variables of gas exchange was not significant, demonstrating that these factors do not interfere jointly in the passion-fruit tree, but there was significance in the isolated factors, Stomatal conductance, intercellular concentration, transpiration and instantaneous water use efficiency were significant for the saline levels, while stomatal conductance, CO_2 assimilation rate and instantaneous water use efficiency differed statistically with the use of silicon (Table 2).

Table 2. Summary of analysis of variance for stomatal conductance (*gs*), intercellular CO_2 concentration (*Ci*), transpiration (*E*), CO_2 assimilation rate (*A*), instantaneous water use efficiency (*EiUA*) and instantaneous carboxylation efficiency (*EiCi*) in 'Gigante Amarelo' passion-fruit grown under different levels of salinity of irrigation water and doses of silicon, at 60 days after transplanting. Pombal, CCTA/UFCG, 2020.

Treatments	GL	QM					
		gs	*Ci*	*E*	*A*	*EiUA*	*EiCi*
Salt levels (NS)	4	0,003**	5365,14**	1,85**	49.24ns	2,61*	0.0001ns
Linear Reg.	1	0,009**	20779,32**	6,01**	25.02ns	9,43**	0.0001ns
Quadratic Reg.	1	0,002*	240.20ns	0.85ns	28,62ns	0.92ns	0.0001ns
Silicon dose (DS)	1	0,002**	846,30ns	0.01ns	3483,44**	3,97*	0.0001ns
Interaction (NSxDS)	4	0.0005ns	826,08ns	0,19ns	147.86ns	0,26ns	0.0003ns
Block	3	0.0013ns	604.66ns	0,85	356,14ns	2.50ns	0.0004ns
Waste	27	0,00054	1125,46	0,27	364,67	0,90	0,00047
CV (%)		28,18	15,57	23,26	19,80	26,26	43,00
				Averages			
Silicon dose							
150 g		0,09 a	220,13 a	2,28 a	87,11 b	3,28 b	0,04 a
300 g		0,07 b	210,93 a	2,25 a	105,78 a	3,92 a	0,05 a

ns, **, * respectively not significant, significant at $p < 0.01$ and $p < 0.05$; Averages followed by different letters indicate significant difference between treatments by Tukey test ($p < 5\%$).

Analyzing the doses of silicon in isolation, it was found that in stomatal conductance (*gs*) the use of the highest concentration of potassium silicate (300 g) reduced *gs by* 22.22% (Table 2), however, it provided an increase in the rate of CO_2 assimilation (*A*) *and* instantaneous water use efficiency (*A/E*), corresponding to 19.15% and 20.0%, respectively. The use of silicon fertilization promotes anatomical alterations in passion-fruit trees, with an increase in the thickness of the adaxial epidermis, reduced palisade parenchyma and an increase in the polar diameter/equatorial diameter of the leaves, which is related to the functionality of the *stomata* (COSTA et al., 2016).

On the other hand, the increment obtained in CO_2 assimilation rate and instantaneous water use efficiency may be related to the deposition of silicon on the leaf wall, which increases the resistance and hardness of the cell walls, reducing cuticular transpiration and consequently increasing water use and photosynthetic efficiency (JESUS et al., 2018).

The *gs* declined linearly as a function of increasing levels of salinity of irrigation water, equivalent to a decrease of 14.57% per unitary increment of ECa, i.e., plants grown under ECa of 3.1 dS m-1 had a reduction in *gs of* 42.67%, in relation to those that were under the lowest saline level of water (0.3 dS m-1) (Figure 1A). Probably, in response to the excess of salts in the root zone, the plant may have reduced the stomatal opening, hoping to reduce leaf transpiration and consequently the absorption of salts, especially Na+ and Cl- (TAIZ et al., 2017). Similar results were found by (FREIRE et al., 2014) in yellow passion-fruit under water salinity, being the research developed in open orchard in the state of Paraíba, constituting the study with irrigation with water of 0.50 and 4.50 dS m-1, where the salt stress inhibited the photosystem II photochemical activity and net photosynthesis of the plants.

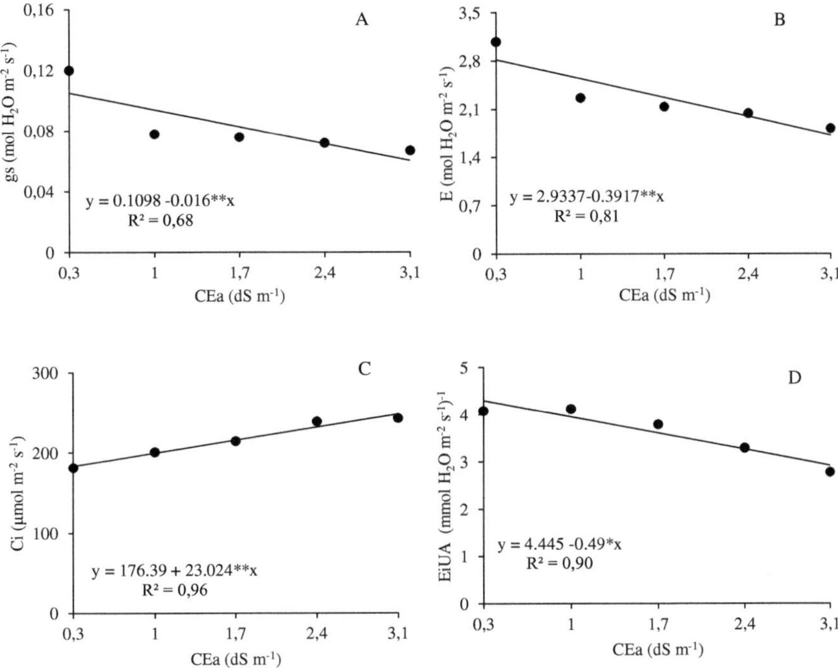

Figure 2. Stomatal conductance - *gs* (A), transpiration - *E* - (B), intercellular concentration *Ci* (C) and water use efficiency - *EiUA* (D) in passion-fruit submitted to salinity in irrigation water and doses of silicon. Pombal, CCTA/UFCG, 2020.

The salinity in the irrigation water also caused a decline in transpiration rate, as well as in instantaneous water use efficiency, with unit losses of 13.35 and 11.02%, respectively (Figures 1B and D). Probably, the passion-fruit plants under salt stress conditions reduced stomatal conductance and transpiration, in order to increase water use efficiency, and with the purpose of avoiding excessive dehydration, but there was also a reduction in *A/E*, perhaps because even under stress conditions the plant continues to photosynthesize normally, maintaining the production of sugars (SILVA et al., 2011; SOUSA et al., 2019).

Irrigation with saline water provided a linear increase in *Ci of* 13.06% per unitary increase of ECa. When comparing the plants submitted to a CEa of 0.3 dS m-1 with those that received the highest level of electrical conductivity of the water, an increase of 247.76 mol H2O m-2 s- was observed, corresponding to an increase of 35.17% (Figure 1C). This fact, reinforces the limitation in the stomatal opening, however it did not restrict the entrance of CO_2 in the cell, because the reduction in conductance, caused a decrease in the absorption of water and salts of the soil solution, without compromising the photosynthetic activity, due to the increase in the concentration of CO_2 may be an indication that the carbon that entered the cell was not assimilated and there was no significant effect on the photosynthesis rate and efficiency of instantaneous carboxylation with water salinity in this study (SILVA et al, 2011; DIAS et al., 2018).

When analysing the photosynthetic pigments by the analysis of variance, it was found that there was no interactive effect, but the levels of water salinity significantly influenced the contents of chlorophyll *a* and carotenoids and the protoplasmic content while the use of silicon significantly influenced chlorophyll *b* and carotenoids (Table 3). Ferraz et al. (2015) in a study carried out at Embrapa Algodão when analysing photosynthetic pigments and cell extrusion in castor oil plant under silicon and salinity where they studied four levels of water salinity (CEa 0, 2, 4 and 6 dS m-1) and four levels of silicon (0, 100, 200, and 300 mg L-1) found significant interactive effect, diverging from this work, in the variable chlorophyll *a* and protoplasmic content.

Table 3. Summary of the analysis of variance for chlorophyll *a* (CL *a*), chlorophyll *b (*CL *b),* carotenoids (CAR) and the protoplasmic content (PC) in passion-fruit 'Gigante amarelo' cultivated under different levels of salinity of the irrigation water and doses of silicon, at 60 days after transplanting. Pombal, CCTA/UFCG, 2020.

	GL	QM			
Treatments		CL *a*	CL *b*	CAR	CP
Salt levels (NS)	4	2,32**	0,45ns	10031,23*	27,92**
Linear Reg.	1	8,77**	1.62ns	37791,94**	105,40**
Quadratic Reg.	1	0.06ns	0.06ns	2292,18ns	0,41ns
Silicon dose (DS)	1	0.01ns	4,11**	77494,56**	12,76ns
Interaction (NS x DS)	4	0,41ns	1.41ns	14671.47ns	1.29ns
Block	3	0,25ns	1,49*	9098,16ns	4.21ns
Waste	27	0,28	0,49	2675,35	3,98
CV (%)		14,97	15,22	10,98	18,07
		Averages			
Silicon dose					
150 g		3,54 a	4,29 b	426,93 b	10,48 a
300 g		3,58 a	4,93 a	514,96 a	11,61 a

ns, **, *, respectively not significant, significant at $p < 0.01$ and $p < 0.05$; Averages followed by different letters indicate significant difference between treatments by Tukey test ($p < 0.05$).

The greatest concentration of chlorophyll *b* and carotenoids in passion-fruit was observed in the plants that were fertilized with the greatest dose of silicon (300 g), with increases in the contents of chlorophyll *b* and carotenoids, of 14.9% and 20.61%, respectively, in the plants that received 300 g in relation to the smallest dose of Si (150 g).

This increased the concentrations of chlorophyll *b* and carotenoids, as well as, the photosynthetic rate of plants (GUERRA et al., 2014). The silicon absorbed by the plants is deposited below the epidermal cuticle, forming a double layer of silica in the cells. This incorporation leads to changes in the architecture of these plants, which keep the leaves more erect, improving the interception of sunlight and therefore photosynthesis.

The synthesis of chlorophyll *a* (Figure 3A) had a linear decreasing behavior as the salinity levels of the irrigation water increased, with a decrease of 10.87% per unitary increment of the ECa. Comparatively, there was a reduction from 4.22 to 2.89 mg g-1 MF, in the interval from 0.3 to 3.1 dS m-1. The reduction in chlorophyll *a* contents according to

Munns & Tester (2008) is the result of imbalances in the physiological and biochemical activities promoted by the salt content in the root zone, beyond that tolerated by the crop, which stimulates the activity of chlorophyllase, an enzyme responsible for the degradation of chlorophyll, inducing the destruction of chloroplasts.

Similar results were found by Wanderley et al. (2018) when evaluating the contents of photosynthetic pigments in passion fruit trees as a function of water salinity in the seedling formation phase, they found a decrease in chlorophyll *a* content of 8.22% for each unit increase in ECa.

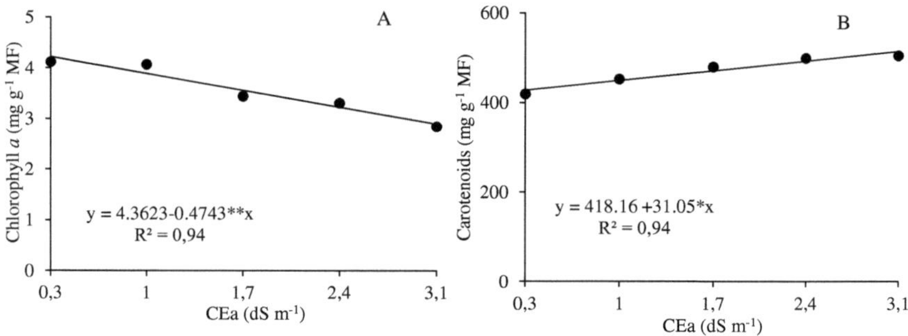

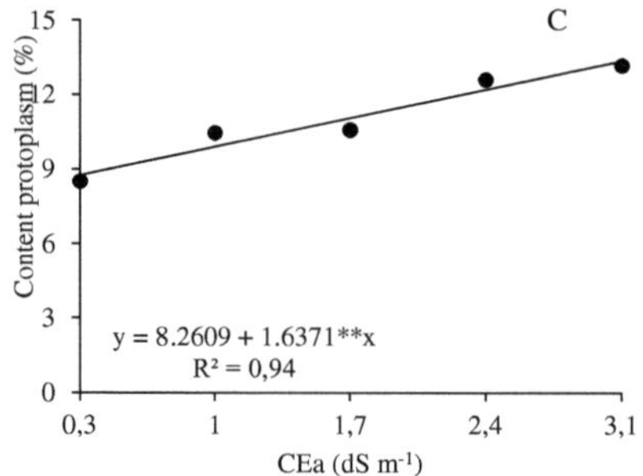

Figure 3. chlorophyll *a* (A), carotenoids (B) and protoplasmic content (C) in passion-fruit tree 'Gigante Amarelo' as a function of electrical conductivity of irrigation water - CEa, at 60 days after transplanting. Pombal, CCTA/UFCG, 2020.

Contrary to what was observed for chlorophyll *a* (Figure 3A), the contents of carotenoids (Figure 3B) increased linearly, with increases of 7.42% per unitary increment of ECa, that is, the passion-fruit plants under EC of 3.1 dS m-1 increased by 20.61% in the synthesis of carotenoids in relation to those that were being irrigated with the lowest level of water salinity (0.3 dS m-1). Carotenoids can act in some situations as antioxidant agents, mitigating the depressive effect of salinity by attenuating photoxidative damage, explaining the increase of this pigment (ASHRAF & HARRIS 2013; LIMA et al., 2017).

The salinity levels of the irrigation water provided an increase in the protoplasmic content of the plants of passion-fruit 'Gigante Amarelo', whose increase was 19.82% per unit increment. Because it is a culture sensitive to the effects of salinity, the increase of salts in the irrigation water caused ruptures in the cell membrane, demonstrated by the gradual increase in the protoplasmic content, indicating release of ions from the cell membrane caused by excess salts (RUTSCHOW et al., 2011). Corroborating the present study, Santos et al. (2018) in research with yellow passion-fruit seedlings under irrigation with salinized water from 0.3 to 4.3 dS m-1 verified an increase of 29% in the protoplasmic content, between the ECa of 0.3 and 4.3 dS m-1.

Salinity also affected the growth of passion-fruit plants, with a significant effect on the stem diameter and the absolute growth rate of the diameter. The doses of silicon significantly influenced the relative growth rate of the stem diameter in the period from 15 to 30 days after transplanting. Similar to what occurred for the physiological variables there was no significant effect of the interaction between the factors (water salinity levels x silicon dose) on the growth of passion-fruit plants (Table 4).

Summary of the analysis of variance for stem diameter (CD), absolute growth rate (ACRdc) and relative growth rate of stem diameter (RGRdc) in passion-fruit 'Gigante Amarelo' subjected to salinity in irrigation water and doses of silicon, 60 days after transplanting. Pombal, CCTA/UFCG, 2020.

	GL	QM		
Treatments		DC	TCAdc	TCRdc
Salt levels (NS)	4	5,87**	0,0016**	0.000007ns
Linear Reg.	1	21,97**	0,005**	0,00002**
Quadratic Reg.	1	0,42ns	0,0008*	0.000001ns
Silicon dose (DS)	1	8,22**	0.000008ns	0,00004**
Interaction (NSx DS)	4	0.56ns	0.003ns	0.000004ns
Block	3	0.72ns	0.0001ns	4.50x 10-7ns
Waste	27	0,46	0,0001	0,000003
CV (%)		5,82	24,94	32,54
			Averages	
Silicon dose				
150 g		11,28 b	0,048 a	0,004 b
300 g		12,19 a	0,047 a	0,006 a

ns, **, * respectively not significant, significant at $p < 0.01$ and $p < 0.05$; Averages followed by different letters indicate significant difference between treatments by Tukey test ($p < 0.05$).

According to the mean test (Table 4), passion-fruit plants fertilized with the highest dose of silicate (300 g) had an increase in CD and TCRdc of 8.07 and 50%, respectively, in relation to those that received the lowest dose of silicate (150 g). As the potassium silicate was used as a source of silicon, for presenting slow release of the mineral, it requires a large quantity to supply the needs of the plants, its use in large quantities provided stem elongation, both because of the benefits of silicon in photosynthesis and consequently growth, and potassium, because it is an enzyme cofactor, osmotic regulator and responsible for meristematic growth (MARSCHNER, 1995). Costa et al. (2016), when studying the action of silicon on the growth of passion-fruit plants, found that this element increased the diameter of the stem up to a concentration of 0.20 g vaseo-1, providing greater plant growth.

The stem diameter and absolute growth rate of diameter reduced 16.39% and 48.40%, respectively, as a function of salt addition in water, independently of the use of silicon (Figures 4A and B). Under salt stress conditions plants reduce growth, due to the consumption of energy for the synthesis of osmotically active organic compounds necessary for compartmentalization processes in the regulation of ion transport (TORRES et al., 2014).

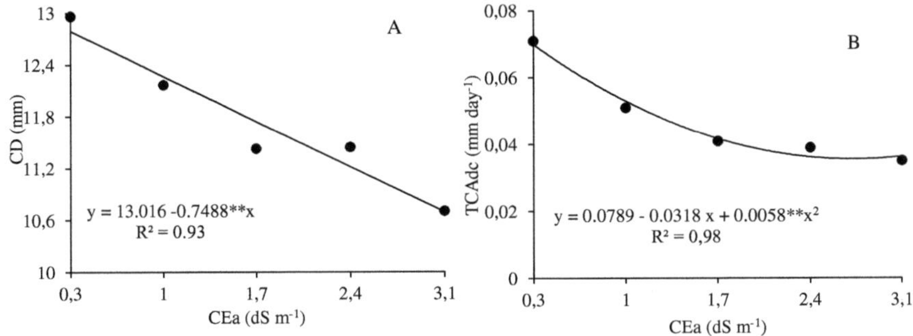

Figure 4: Stem diameter (A) and absolute growth rate (B) in passion-fruit tree 'Gigante Amarelo' as a function of electrical conductivity of irrigation water - CEa, at 60 days after transplanting. Pombal, CCTA/UFCG, 2020.

Bezerra et al. (2016) also observed reductions in the growth rate of passion-fruit plants as a function of salinity in water. Andrade et al. (2018) observed a 31% decrease in stem diameter in passion-fruit seedlings irrigated with water containing the lowest (0.2 dS m-1) and highest (6.2 dS m-1) salt contents in irrigation water.

4 CONCLUSIONS

Salt stress caused reductions in gas exchange, synthesis of chlorophyll *a* and *b and* in the growth of passion-fruit plants at 60 days after transplanting.

The use of the highest silicon concentration (300 g) provided increases in CO_2 assimilation rate and instantaneous water use efficiency, corresponding to 19.15 and 20.0%.

Passion-fruit plants fertilized with the highest dose of silicon (300 g) had an increase in CD and TCRdc of 8.07% and 50%, respectively, in relation to those that received the lowest dose of silicon (150 g).

5. BIBLIOGRAPHIC REFERENCES

ANDRADE, J.R.; SOUSA, A.; JUNIOR, M.; OLIVEIRA, S.; PAULA; REZENDE, L. P.; ARAÚJO NETO, J. C. Germination and morphophysiology of passion fruit seedlings under salt water irrigation. **Tropical Agricultural Research**, v. 48, n. 3, p. 229-236, 2018.

AMORIM, A. V.; GOMES-FILHO, E.; BEZERRA, M. A.; PRISCO, J. T.; LACERDA, C. F. Physiological responses of adult early dwarf cashew plants to salinity. **Revista Ciência Agronômica**, v.41, p.113-121, 2010.

ASHRAF, M.; HARRIS, P.J.C. Photosynthesis under stressful environments: An overview. **Photosynthetica,** v.51, n.2, p.163-190, 2013.

ARNON, D. I. Copper enzymes in isolated chloroplasts: polyphenoloxidases in *Beta vulgaris*. **Plant Physiology**, v.24, n.1, p.1-15, 1949.

BEZERRA, J. D.; PEREIRA, W. E.; SILVA, J. M.; RAPOSO, R. W. C. Crescimento de dois genótipos de maracujazeiro-amarelo sob condições de Salinidade. **Revista Ceres**, v.63, n.4, p. 502-508, 2016.

BENINCASA, M. M. P. Análise de crescimento de plantas: notções básicas. Jaboticabal: **FUNEP**, 2003. 42 p.

BERNARDO, S; SOARES, A. A.; Mantovani, E. C. **Manual de irrigação**. Viçosa: UFV, 2006.625p.

BERTOLLI, S. C.; SOUZA, J.; SOUZA, G. M. Photosynthetic characterization of the isohydric species pata-de-elefante under water deficiency conditions. **Revista Caatinga**, v.28, p.196-205, 2015.

CAMPOS, P. S.; THI, A. T. P. Effect of abscisic acid pretreatment on membrane leakage and lipid composition of *Vigna unguiculata* leaf discs subjected to ormotic stress. **Plant Science**, v. 130, n. 1, p.11-18, 1997.

CAVALCANTI, J.C.P. Recomendações de adubação para o estado Pernambuco (2ªaproximação). 3.ed. Recife: **Instituto agronômico do Pernambuco - IPA**, 2008, 212p.

COSTA, B. N. S.; COSTA, I. J. S.; DIAS, G. M. G.; ASSIS, F. A.; PIO, L. A. S; SOARES, J. D. R.; PASQUAL, M. Morpho-anatomical and physiological alterations of passion fruit fertilized with silicone. **Pesquisa Agropecuária Brasileira**, v.53, n.2, p. 163-171, 2016.

CLAESSEN, M. E. C. (Org.). Manual de métodos de análise de solo. 2. ed. rev. atual. Rio de Janeiro: **Embrapa-CNPS**, p.212, 1997.

DIAS, A. S.; LIMA, G. S.; SÁ, F. V. S.; GHEYI, H. R.; SOARES, L. A.A.; FERNANDES, P. D. Gas exchanges and photochemical efficiency of West Indian cherry cultivated with saline water and potassium fertilization. **Brazilian Journal of Agricultural and Environmental Engineering**, v.22, n.9, p.628-633, 2018.

DIAS, N. S.; BLANCO, F. F.; SOUZA, E. R.; FERREIRA, J. F.; SOUSA NETO, O. N.; QUEIROZ, I. S. R. Efeitos dos salais na planta e tolerância das culturas à salinidade. In: GHEYI, H. R.; DIAS, N. S.; LACERDA, C. F.; GOMES FILHO, E. **Manejo da salinidade na agricultura: Estudo básico e aplicados.** 2ª ed. Fortaleza: INCTSal, 2016. cap.11, p.151- 161.

FERREIRA, D. F. SISVAR: A computer statistical analysis system. **Ciência e Agrotecnologia**, v. 35, n. 06, p. 1039 - 1042, 2011.

FERRAZ, R. L. S.; MAGALHÃES, I. D.; BELTRÃO, N. E. M.; MELO, A. S.; BRITO NETO, J. F.; ROCHA, M. S. Photosynthetic pigments, cell extrusion and relative leaf water content of the castor bean under silicon and salinity. **Revista Brasileira de Engenharia Agrícola e Ambiental**, v.19, n.9, p.841-848, 2015.

FREIRE, J. L. O.; DIAS, T. J.; CAVALCANTE, L. F.; FERNANDES, P. D.; NETO LIMA, A. J. Quantum yield and gas exchange in yellow passion fruit under water salinity, biofertilization and mulching. **Revista Ciência Agronômica**, v.45, n.1, p. 82-91, 2014.

GUERRA, A. M. N. M. RODRIGUES, F. A.; LIMA, T. C.; BERGER, P. G.; BARROS, A. F.; SILVA, Y. C. R. Photosynthetic capacity of cotton plants infected by ramulose and supplied with silicon. **Bragantia,** v.73, n.1, p.50-64, 2014.

GHEYI, H. R.; DIAS, N. S.; LACERDA, C. F. Manejo da salinidade na agricultura: Estudos básicos e aplicadas. **INCT Sal**, 2010. 472p.

GONG, H.; ZHU, X.; CHEN, K.; WANG, S.; ZHANG, C. Silicon alleviates oxidative damage of wheat plants in pots under drought. **Plant Science**, v. 169, p. 313-321, 2005.

HATTORI, T.; INANAGA, S.; ARAKI, H.; AN, P.; MORITA, S.; LUXOVÁ, M.; LUX, A. Application of silicon enhanced drought tolerance in *Sorghum bicolor*. **Physiologia Plantarum**, v. 123, n. 4, p. 459-466, 2005.

JESUS, E. G.; FATIMA, R. T.; GUERRERO, A. C.; ARAÚJO, J. L.; BRITO, M. E. B. Growth and gas exchanges of arugula plants under silicon fertilization and water restriction. **Brazilian Journal of Agricultural and Environmental Engineering,** v.22, n.2, p.119-124,2018.

LIMA, M. A.; CASTRO, V. F.; VIDAL, J. B. ENEASFILHO, J. Aplicação de silício em milho e feijão-de-corda sob estresse salino. **Revista Ciência Agronômica**, v. 42, n. 2, p. 398-403, 2011.

LIMA, G. S.; DIAS, A. S.; GHEYI, H. R.; SOARES, L.A.A. ; NOBRE, R. G. ; SÁ, F. V. S. ; PAIIVA, E. P. Emergence, morpho-physiology and flowering of colored-fiber cotton (*Gossypium hirsutum* L.) submitted to different nitrogen levels and saline water stress irrigation. **Australian journal of crop science**, v. 11, n. 7, p. 897-905, 2017.

MARSCHNER, H. Mineral nutrition of higher plants. San Diego: **Academic Press**, 1995. 889 p.

MUNNS, R.; TESTER M. Mechanisms of salinity tolerance. **Annual Review of Plant Biology**, v. 59, p. 651-681, 2008.

RICHARDS, L. A. Diagnosis and improvement of saline and alkali soils. Agriculture Handbook No. 60, Washington: USDA, Department of Agriculture, 1954. 160 p.

RUTSCHOW, H. L.; BASKIN, T. I.; KRAMER, E. M. Regulation of solute flux through plasmodesmata in the root meristem. **Plant Physiology,** v.155, n. 1, p.1817-1826, 2011.

SANTOS, J. B. Estudo das relações nitrogênio:potássio e calcio:magnésio sobre o desenvolvimento vegetativo e produtivo do maracujazeiro amarelo. 2001. 88 f. Dissertation (Master in Soil and Water Management). Centre for Agrarian Sciences, Federal University of Paraíba.

SANTOS, A. S.; ARAÚJO, R. H. C. R.; NOBRE, R. G.; SOUSA, V. F. O. RODRIGUES, M. H. B. S.; FORMIGA, J. A.; GOMES, F. A. L.; SANTOS, G. L.; ONIAS, E. A. Effect of hydrogen peroxide in the growth of yellow passion fruit seedlings under salinity stress. **Journal of Agricultural Science**, v.10, n.10, p.151-162, 2018.

SILVA, E. N.; SILVEIRA, J. A. G.; FERNANDES, C. R. R.; DUTRA, A. T. B.; ARAGÃO, R. M. Acúmulo de íons e crescimento de pinhão-manso sob diferentes níveis de salinidade. **Revista Ciência Agronômica**, v.40, p.240-246, 2009.

SILVA, E. N.; RIBEIRO, R. V.; SILVA, F. S. L.; VIÉGAS, R. A.; SILVEIRA, J. A. G. Salt stress induced damages on the photosynthesis of physic nut young plants. **Scientia Agrícola**, v.68, n. 1, p.62-68, 2011.

SOUSA, V. F. O.; COSTA, C. C.; DINIZ, G. L.; SANTOS, J. B.; BONFIM, M. P.; LOPES, K. P. Growth and gas changes of melon seedlings submitted to water salinity. **Revista Brasileira de Engenharia Agrícola e Ambiental,** v.23, n.2, p. .90-96, 2019.

TAIZ, L.; MOLLER, E. Z. I. M.; MURPHY, A. Fisiologia e desenvolvimento vegetal. 6.ed. **Artmed**, 2017. 918p.

TAIZ, L.; ZEIGER, E. Plant physiology. 4nd ed. Sunderland: Sinauer Associates, Inc. Publishers, 2009. 848p.

TORRES, E.C.M.; FREIRE, J. L.M.O.; OLIVEIRA, J. L.; BANDEIRA, L. B. Biometria de mudas de cajueiro anão irrigadas com águas salinas e uso de atenuadores do estresse salino. **Nativa**, v.2, n.2, p.71-78, 2014.

WANDERLEY, J.A.C.; AZEVEDO, C. A. V.; BRITO, M. E. B.; CORDÃO, M. A.; LIMA, R. F.; FERREIRA, F. N. Nitrogen fertilization to attenuate the damages caused by salinity on yellow passion fruit seedlings. **Brazilian Journal of Agricultural and Environmental Engineering**, v.22, n.8, p.541-546, 2018.

MIX
Papier aus verantwortungsvollen Quellen
Paper from responsible sources
FSC® C105338

Printed by Books on Demand GmbH, Norderstedt / Germany